Putting "Loafing Streams" *to Work*

PUTTING
"Loafing Streams"
TO WORK

THE   BUILDING   OF

**LAY, MITCHELL, MARTIN, AND JORDAN DAMS,
1 9 1 0 - 1 9 2 9**

HARVEY H. JACKSON III

THE UNIVERSITY OF ALABAMA PRESS

Tuscaloosa and London

∞

The paper on which this book is printed meets the minimum requirements of
American National Standard for Information Science–Permanence of Paper for
Printed Library Materials, ANSI Z39.48-1984.

Library of Congress Cataloging-in-Publication Data

Jackson, Harvey H.
Putting "loafing streams" to work : the building of Lay, Mitchell,
Martin, and Jordan Dams, 1910–1929 / Harvey H. Jackson, III.
p.   cm.
Includes bibliographical references and index.
ISBN 0-8173-0879-2 (cloth : alk. paper).—ISBN 0-8173-0889-x
(pbk. : alk. paper)
1. Dams—Alabama—History.   2. Alabama Power Company—History.
3. Construction workers—Alabama—History.   I. Title.
TC557.A2J33   1997
627'.8'09761—dc21   96-48810

British Library Cataloguing-in-Publication Data available

Cover illustrations: Jordan Dam and (inset) fishermen at Cherokee Bluffs,
shortly before work on Martin Dam began. (Alabama Power Company Archives)

For my Alabama history teachers

Contents

Acknowledgments

I was working on another book when I met Bill Tharpe. Tharpe is the Alabama Power Company's archivist, and I went to him to gather information about hydroelectric development of the Coosa and Tallapoosa. Among the first things he showed me were the rich resources available on the company's early dams, and it did not take me long to confirm what Bill already knew—here was a story that needed to be told.

I finished my research, went away, and wrote *Rivers of History: Life on the Coosa, Tallapoosa, Cahaba, and Alabama.* But the dams were only a small part of that book, and I wanted to do more. Before long I was back in touch with Tharpe, and we began discussing how we might pull it off. Encouragement from Senior Vice President Robert Buettner and Manager of Corporate Information Dave Rickey was essential at this early stage, and so it followed quickly that I put together a proposal, the company agreed to my plan, and work got underway.

Records in the Alabama Power Company's archives obviously were the key to my research, and the support I received there was essential to the project. But equally important was the help I received in getting in touch with people who had direct knowledge of how the dams were built and of the people who built them. Requests went out to various company offices, and employees responded with names and addresses of people I should contact. Other sources appeared when newspapers that serve communities close to the dams ran stories about my research, and soon I discovered a network of "dam people" who were as interested and excited about the book as I was. Before I finished, I had met some fascinating folks and made many new friends.

Although most of my contacts are mentioned in the end notes, I want to recognize those who were a special help. Frank D. Greene, Fred G. Mayfield, John D. Glascock, and Geraldine Hollis (Mrs. Harold Lawrence) took time to talk with me about Lay and Mitchell dams and to help me understand

x what those projects meant to the lives of people who lived in the area. Over on the Tallapoosa a host of kind souls appeared to aid me, and to all of them—Ben Hyde, L. B. "Gip" O'Daniel, Judge C. J. Coley, John D. Towns, Beula Golden Ingram, Gordon Gauntt, L. B. Crouch, Lloyd Frank Emfinger, Dr. J. F. Fargason, Clyde Steverson, and Guesna Neighbors Moon—I give my sincere thanks. B. K. McDonald, Marguerite "Bill" Henry Roach, Howard F. Bryan III, L. B. Groover, Jr., and Sarah Cabot Robison Pierce also took time to talk with me and answer the many questions I had about the dams and the people who built them. Letters and telephone calls from Joyce Rabe, Sue Comerford, Frank F. Finn, Charles Reynolds, Frank M. Dunlap, Elizabeth Duncan Crowder, Larry Lee Waites, Harold Hill, Richard S. Woodruff, Emma Frank Bowers, Roger Townley, Jim Cassidy, J. Orin Hardin, and Charles Adair were also greatly appreciated, as was the help from newspaper writers Nick Lackeos, Michael R. Kelley, Jack B. Venable, Melanie Jones, Gerald Williams, and Lewis Scarborough. Back at Alabama Power Charles "Buddy" Eiland and Chris Conway read the manuscript, caught errors, and offered suggestions. Without the support and encouragement of all these people, things would not have gone as well as they did.

Jacksonville State University and my colleagues in the history department also helped. The university, and especially Dean Earl Wade, allowed me to rearrange my schedule so I could have blocks of free time to complete this project. Ted Childress, Phil Koerper, Suzanne Marshall, Buddy Hollis, and Ron Caldwell, each in their turn, took up department head duties when I was away. Linda Cain of the Houston Cole Library helped me identify sources. Audrey Smelley and Reginna Horne gave the manuscript a critical reading and offered suggestions. Also important was the help given me by the staff of the Chilton-Clanton Public Library, the Tallapoosa County Court House, and the Alabama Department of Archives and History. Leah Rawls Atkins went beyond the call of friendship and duty when she read and commented on the manuscript, and my work is much improved as a result. I also appreciate the help of Marlene Hunt Rikard, who gave me the benefit of her knowledge of industrial development during the period. And my sincere thanks goes to the folks at The University of Alabama Press—especially to Malcolm MacDonald who got the book started; to Nicole Mitchell who saw it through to the end; and to Suzette Griffith who was with it all the way.

I want to thank my father, Harvey H. Jackson, Jr., and my aunt, Anne Jackson Bennett, who both grew up on Lake Jordan and who had their own tales to tell. My wife, Suzanne, supported me with kind words, displayed a tolerant forgiveness for those times I went off with my tape recorder, and gave me all the love I needed. My daughter, Kelly, grown up and gone away, encouraged me as well, while my son, Will, kept me firmly grounded in the reality and the joy of fatherhood. His further contribution was to sleep late so I could work early and to take naps in the afternoon while I wrote—just as a two-year-old should do.

Last I want to acknowledge those good folks who over the years taught me in and of Alabama. In Grove Hill, where I grew up, Mrs. Willie Tucker introduced me to our state's past when I was in the fourth grade, and I still remember what and how she taught. Miss Nannie Rae Wilson continued my education in junior high, and Mr. Melvin Joiner kept up my interest in history in senior high, despite competition from raging hormones. When I got to Marion Institute, Col. James Jackson showed me what a college history teacher was and made me want to be one; then at Birmingham Southern Dr. O. Lawrence Burnett helped me realize that I wanted to be a historian as well. At The University of Alabama Drs. Thomas B. Alexander, John Pancake, Charles Summersell, and James Doster rounded out my education in the state. Then I left and did not return for twenty-five years. But because of them Alabama was always with me, and to them I dedicate this book.

Putting "Loafing Streams" *to Work*

The Lower Coosa and Lower Tallapoosa

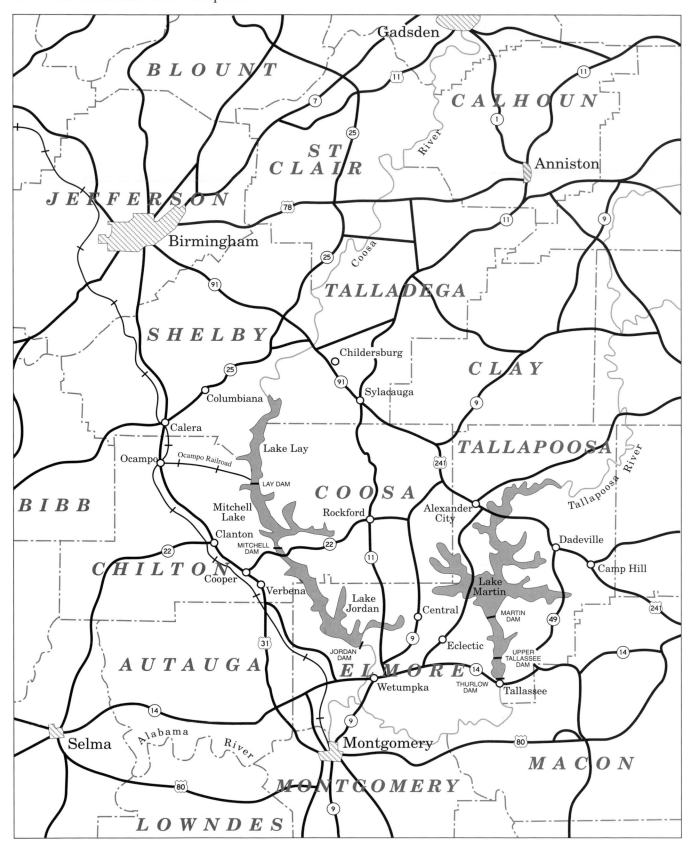

In the Beginning

I now commit to you the good name and destiny of the Alabama Power Company. May it be developed for the service of Alabama.

—WILLIAM PATRICK LAY to JAMES MITCHELL
and THOMAS W. MARTIN, May 1, 1912

BETWEEN 1912 AND 1928 THE ALABAMA POWER Company built four hydroelectric dams on the Coosa and Tallapoosa Rivers. At the time, and for some time to come, these dams were the largest, most complex, and most expensive construction projects in the history of the state. But they were more than that. The dams were a turning point. When their turbines began spinning and electricity began to flow from the powerhouses to farms, towns, and cities, Alabama moved from the nineteenth into the twentieth century. The state and its people would never be the same again.

The origins of the Alabama Power Company have become part of Alabama folklore. The story begins with Capt. William Patrick Lay, scion of three generations of river men and a pioneer in power production in the northeastern part of the state. According to tradition, one day as Lay watched water rush through the flume at Lock 2 on the Coosa, it came to him that a dam across the river could harness the stream and generate electricity for the piedmont. That dream refused to die, and in 1906 he founded the Alabama Power Company to make the dream come true. In the meantime the captain looked about for sites. He soon found the one he wanted, a remote spot on the Coosa River, some eleven miles from Clanton, that was known to locals as the site for the proposed Lock 12. There, he told the congressional committee that approved his application, he would build a plant

William Patrick Lay at the site of the
future Mitchell Dam, Coosa River, 1922

(Alabama Power Company Archives)

A Gentleman who, for more than half a century has steadfastly pointed out the value of the development of Alabama's internal waterways and hydro-electric power.

that would generate electricity on a scale greater than the region had ever known. Congressional approval alone was not enough, however. Lay had promised to build the dam with private money, but he soon found that northern capitalists (the only ones with sufficient funds to back him) had little faith in the project. Things did not look good, and by 1911 Lay was near giving it up when he met two men: Canadian-born, Massachusetts raised, and internationally connected James Mitchell and his associate, a young Montgomery lawyer of an old, respected family, a man who knew Alabama laws, customs, and conditions, Thomas W. Martin.[1]

Like Lay, today Martin and Mitchell seem almost mythical characters. In

1911, when they began to combine their talents, Mitchell was forty-five years old, fifteen years Martin's senior. Behind James Mitchell were over two decades of work for power companies in America, Brazil, and Japan that had made him, according to associates, "one of the most experienced men in the world of electrical power." Martin, for his part, already represented groups interested in the power potential of the Tallapoosa River, so when Mitchell arrived looking for advice on hydroelectric possibilities in Alabama, he soon found himself at the attorney's office. A short time later the two men were in the field, exploring rivers and assessing sites; the best, they agreed, was Cherokee Bluffs on the Tallapoosa, upstream from Tallassee. But Mitchell's vision was of more than a single undertaking. In his plan for the future, Cherokee Bluffs would be only one of a series of similar projects, all part of a comprehensive plan for exploiting the hydroelectric potential of both the Coosa and the Tallapoosa.[2]

Mitchell brought to Alabama more than extensive experience and grand plans. Most important of all, he arrived with connections to "old banking friends" in England and Canada who would share his vision and invest in it. So while Lay's efforts to raise funds floundered, Mitchell's were successful. Then, with money in hand (or at least in the bank), he and Thomas Martin opened talks with individuals and interests that owned or controlled the sites where dams could be built and held title to the land that would be flooded when the gates were closed. Their first choice remained Cherokee Bluffs, so Martin focused most of his attention on the Tallapoosa acquisition. Among the other properties they needed to control if Mitchell's plans for the comprehensive and coordinated development of the rivers would ever become reality was Lock 12 on the Coosa. That site was held by William Patrick Lay and his Alabama Power Company.[3]

Martin and Mitchell became aware of Alabama Power's holdings at Lock 12 at about the same time that the New York investors who were Lay's last hope for financing withdrew from the project. At that point, as Lay later put it, to his "relief and satisfaction . . . James Mitchell, representing himself and clean, high-minded English and Canadian capitalists . . . came to our rescue." What Captain Lay called a "rescue" might today be called a "takeover," though decidedly not a hostile one. In the negotiations that followed, Martin and Mitchell made Lay an offer, and Lay accepted. The papers were drawn

James Mitchell, c. 1915
(Alabama Power Company Archives)

Thomas W. Martin, 1910, in his Montgomery
office. The moustache did not last out the year.

(Alabama Power Company Archives)

up, and on May 1, in Thomas Martin's Montgomery law office, William
Patrick Lay placed "the good name and destiny of the Alabama Power Company" into the hands of James Mitchell and Thomas W. Martin. In the years
to come Lay would serve the company in a variety of roles, but he was no
longer its leader. The torch had passed.[4]

The wisdom of gaining control of Alabama Power and its assets was soon
apparent. Still planning to develop Cherokee Bluffs first, Martin moved the

project through the tangle of local, state, and national rules and requirements that the company faced. As he did his efforts came into conflict with downstream interests whose opposition threatened to tie the company up in court for months, if not years. Now, faced with pressure from investors to begin soon and encountering legal obstacles that kept them from starting at the Bluffs, the managers of the new Alabama Power Company shifted their focus and decided that Lock 12, which was acquired when they obtained the name, would be the site of the company's first hydroelectric dam.[5]

The arrangement with Lay included rights to the property *and* congressional approval to build the dam. It also included some restrictions. In 1906 the federal government told Lay that work at Lock 12 must begin by 1909. A bit of preliminary construction Lay ordered at the site met this requirement in spirit, if not in substance; but the authorization also required that the dam be completed by March of 1914. Now Lay's deadline, like the company, belonged to Mitchell and Martin.[6]

The deadline was not theirs alone. In short order they imposed it on Eugene A. Yates, the engineer brought to Alabama to design and oversee the Lock 12 project and on Oscar G. Thurlow, whom Yates employed as his assistant. Alabama Power also set that goal for the MacArthur Brothers Company of New York, the contractors hired to actually construct the dam. And so it followed that March 1914 became the deadline for the hundreds of workers hired to build roads, clear land, blast rock, build forms, pour concrete, hang lines, run machines, cook food, supervise labor, and keep order in what would become, for a short time, the largest town between Montgomery and Birmingham. They were all in it together. Success or failure depended more on some than on others, but in the scheme of things, everyone was important.[7]

What follows is the story of how Alabama Power built the dam at Lock 12 before World War I, then constructed three more dams in the 1920s, before the Great Depression brought everything to a halt. Although to some extent this story has been told before, the account laid out here is fundamentally different from previous histories of the company's early days. Past chronicles were written by Alabama Power insiders, managers, and executives, including Thomas W. Martin himself. Their focus was on what corporate leadership did to guide the company through those first few decades.

Lock 2 on the Coosa, the spot where
William Patrick Lay supposedly got the idea for
generating hydroelectric power on the river

(Scarboro Photography, Gadsden, Alabama)

This study focuses instead on the people who actually built the dams, how they
worked and how they played, the dangers they faced, and the impact their
presence had on residents who lived in the immediate area. This is the story
of workers clearing thousands of acres of wilderness, changing the course of
mighty streams, pouring tons of concrete, raising dams from the river beds,
carrying their injured from the site, and, at times, burying their dead.[8]

Despite a different emphasis, this is still the story of the Alabama Power
Company. Although decisions made in the boardroom in Birmingham will

8 be considered only as they affected what happened at the sites, in those early days executives often took note of such mundane matters as outhouses, ice and water, flies and mosquitoes, cold drinks in the commissary, pool rooms, and Sunday schools. At times it seemed that no detail was too small, but Martin and Mitchell understood the risk they were taking and knew that details could doom a project. Today it is easy to forget that at the outset Alabama Power was an unproven company that had neither generated nor sold a single kilowatt. It is also easy to forget that there were some Alabamians who wondered if it ever would.

Looking back, one can see why the skeptics had their doubts. Although the company would rely on men experienced in hydroelectric development, the industry was still in its infancy, and many of its problems were yet to be solved. Moreover, the very things that made Lock 12 and the other locations so desirable also made construction more difficult and dangerous. The sites' advantages included steep banks to anchor the dam and hold the lake, bedrock for a foundation, acres of trees for timbers and boards, and a small, scattered population, so that few farms would be flooded and have to be bought at premium prices. Indeed, when Lay sought congressional approval for a project at Lock 12, one of his strongest arguments was that if the dam failed and there was a washout, no one would be hurt because no one lived anywhere in the vicinity. But the rough terrain that kept people out and kept the land undeveloped and cheap meant that the few roads in the region were little more than trails, that bridges were makeshift at best, and that railroads—well, there were none. All of these would have to be built just to get men and machinery to the site for work to begin.[9]

Access to the river was just the first of many problems that had to be overcome. One of the most daunting tasks facing contractors was finding men to do the work. Most of the people who lived near the dams were, in one way or another, farmers; and those who were not depended on the farm economy to provide customers for their stores, shops, and other businesses. Only a few—a carpenter here and there, and perhaps a mason—possessed the skills that would be needed to build hydroelectric dams. If the job was to be done and the deadline met, company contractors had to bring in more carpenters and masons, plus machinists, electricians, heavy equipment operators, and a host of other skilled workers. But more than skilled workmen were lacking. Common laborers by the hundreds were needed, and a force of that size could not be drawn from the immediate area. Once hired, these workers, skilled and unskilled, had to construct the camps where they would be housed and fed, clear the land where the lakes would rise, survey and grade the roads, lay the rails, build and fill storage facilities, blast the quarries, crush the rock, and locate sand and gravel. Moreover, most of these things had to be done before a yard of concrete could be poured.[10]

The physical effort involved was no more important and no more complex than the social scope of the projects, for dam builders had to overcome problems of personnel management that few in Alabama had ever faced—at least on such a scale. Not only did laborers have to be recruited and brought to the remote sites, but they also had to be acculturated to living and working conditions that were new and strange to most. The company's expectations were simple enough: employees were supposed to work hard, follow orders, and behave. In return they would be treated fairly and paid as well as the profit margin allowed. Consequently, management adopted policies that reflected a concept that was popular among many industrial leaders at the time—welfare capitalism. Defined as a service or services "provided for the comfort or improvement of employees which was neither a necessity of the industry or required by law," welfare capitalism seemed to offer employers a way to keep workers productive on the job and out of trouble when their shift was over. According to advocates, if employees were adequately housed and fed, if health and sanitation were company priorities, and if there was wholesome entertainment for idle hours, there would be less malingering, less absenteeism, less drunkenness, and of course, less job turnover. There would also be less reason for workers to join unions, a goal some historians contend was the main purpose of welfare capitalism. But unions were of little concern to Alabama Power on those early projects. What the company wanted instead was a dependable workforce, and the ideas advanced by welfare capitalists seemed a reasonable way to achieve that end. Martin, Mitchell, and their associates were practical men, and their labor-management policies reflected that outlook.[11]

Not surprisingly, Alabama Power's personnel policies also reflected certain class assumptions. Believing that most members of the working class would not (or could not) control themselves without supervision, Alabama

Power's middle-class managers drew up personnel guidelines designed to regulate behavior and channel excess energies into productive activities. What company leaders hoped for was a paternalistic relationship with responsible, professional supervisors guiding and protecting the workers under their charge. Workers often had other ideas. Many chafed under these restrictions, and in some cases they forced the company to change its rules. What resulted, after four dams' worth of revision and recalculations, was an arrangement between management and labor that was satisfactory to those concerned. How this system came into being is an important part of the story, for its evolution reveals much about the attitudes and expectations of everyone involved.[12]

Class assumptions were only part of the equation. There was also race. This was Alabama early in the twentieth century. White supremacy was the dominant racial theory, Jim Crow laws formally separated the races, and politicians ran and won on the promise that they would keep the "Negro in his place." Blacks who dared violate the rules that defined their status quickly felt the wrath of the white community. Local newspapers regularly reported such instances, and in 1914, the year that the dam at Lock 12 went on line, a black man accused of "brutally assaulting" a white woman was chased down west of the nearby town of Clanton and his "body suspended by a rope from a tree" while members of the mob "riddled [it] with bullets." As usual, no one was arrested, for whites dominated local law enforcement just as they did the local economy. Although blacks throughout the South had a long history of working around the system and negotiating with whites to ease the burdens of some of its most offensive elements, there were still times, like this one, when negotiation and intercessions failed, and blacks had little choice but to accept things as they were and make the best of a bad situation.[13]

Alabama Power Company, like every other white-owned and -operated industry in Alabama, had no intention of altering this racial arrangement. Like their counterparts throughout the South, Alabama Power's management believed that blacks were even less capable and less responsible than working-class whites, and therefore African Americans needed to be regulated even more closely. This conviction was reflected over and over in the treatment of black workers both on and off the job, and today records of what occurred offer us another look at how racial attitudes and assumptions shaped the lives of Alabamians early in this century. But even though the Alabama Power Company did not come on the scene to bring about social change, its arrival introduced a new factor into the racial equation and threatened to undermine the foundation on which local white control was built. The most obvious source of common labor for construction was the local black community, and local blacks were excited over this new employment opportunity. Whites, especially white landlords, were less enthusiastic. They feared that a job at a dam might take their sharecroppers and tenants away from the farm during those critical planting, chopping, and picking months, when the success or failure of the cotton crop was determined. To get them back landlords might have to offer higher wages, better crop divisions, or other incentives, which would reduce profits and, more important loosen the hold that "the man" had over "his" workers. A dam offered blacks an alternative, and an alternative was not what those in authority wanted blacks to have.[14]

Even if Alabama Power recruited its black laborers elsewhere, brought them to the site, and housed them in the company camp, local whites were still concerned. The idea of a large group of "foreign" blacks living together in the region made many uneasy, and the company was hard put to assure whites that such a settlement was not a threat to the greater community. To calm those fears Alabama Power, following the example set by the Birmingham steel industry, went to great lengths to keep the races separate and make sure that blacks lived and worked under white supervision. Anything less, and local support for the project would have evaporated. But solving this problem threatened to create another. How could the company get blacks to hire on at the project if all it offered was the most menial jobs for the lowest wage, the poorest houses, strictly segregated conditions, and close control by whites? What inducements could Alabama Power give to get men to accept such employment? The answer: steady work and regular pay, two things denied many African Americans. And since the other conditions were no worse than those they faced just living in Dixie, they signed on. Some did not like the job and soon quit. Others, however, stayed, worked for and with the company, and became important (though secondary) members of the team. In time, with Alabama Power's cooperation (or at least its acquies-

Alabama Power officials and their guests
inspect the Lock 12 site.

(Alabama Power Company Archives)

cence), black workers were able to amend some rules and alter some conditions to make things a little better for themselves and for their families.[15]

White concerns over the impact these Alabama Power projects might
have on blacks, local and otherwise, were offset by the benefits the projects
promised to the local economy and to the white community at large. In the
short term there would be jobs, money, and purchases. Local businessmen
saw golden opportunities, and many seized them. Dam building ushered in
the flush times, and even though the boom was soon over, it was universally
enjoyed. But Alabama Power brought with it hope for more than just a few

good years. It is no exaggeration to say that when the dams were built, the Coosa-Tallapoosa valley was more a part of the nineteenth century than the twentieth. Farmers worked the land much as their fathers and grandfathers had done, with mule, plow, sweat, and muscle. Even in the towns wood stoves were still the rule, kerosene lamps lit most homes, automobiles were few, and the local blacksmith shod horses and repaired buggies just as always. Occasional visits to Montgomery, Selma, or Birmingham exposed some citizens to modern conveniences while showing them just how far they had to go before they could enjoy the same. Such trips also revealed that the key to progress, the key that would open the door to the modern world, was electricity. The arrival of the Alabama Power Company meant, locals reasoned, that power would soon be available to them at rates they could afford. And once they had that power, nothing could hold them back.

Yet we must not forget that for Alabama Power, success did not always come easily. In many respects the company had to learn on the job, for each project presented new and often unforeseen challenges. Every undertaking involved experimentation and learning, and ideas that did not work at one site were seldom tried out at another. As a result, the men of Alabama Power—those at the dam and those at headquarters—carefully assessed each project when it was done, passed judgment on practices and policies, and set out to do things better the next time. They also set down their observations for others to consult, and others did I was one of them, and without their comments I doubt if this book could have been written.

In the final analysis, what Alabama Power did on the Coosa and Tallapoosa involved more than building dams and harnessing rivers. It involved getting men to share a dream and work together to make that dream a reality. Because some among them dreamed different dreams and some cared little about working with people who were not like themselves, the task was harder than it might have been. But the job was done, four times, and the dams still stand as monuments to the effort.

A Dam at Lock 12

*When Capt. Lay first began to come to Clanton and talk about damming
up the Coosa River people laughed at his theory. . . . They said that fellow
knows better than such as that. He is just talking that stuff to keep us from
knowing what he is really up to. He is some crank who is hunting for a gold
mine out here in these hills, and just talks about damming up the river to
keep people from catching on to his foolish mission here.*

—*Clanton Union Banner,* November 24, 1921

FOLKS IN CLANTON WERE NOT QUITE SURE WHAT
to make of the slight, energetic, middle-aged man who appeared in
their midst and introduced himself as William Patrick Lay. He did
not linger long, just paused to get directions, then headed toward
the Coosa. Soon word drifted back that in the man's luggage was a sack of
fifty-dollar gold pieces and that a farmer could have one if he would give
up forty acres near the river. To folks who asked him why anyone would
want scrubby bottoms and hollows out there in the middle of nowhere, Captain Lay explained how he planned to build a hydroelectric dam across the
stream, down at Lock 12, and that he was buying up what would be flooded.
Now this all seemed pretty far-fetched to locals. A dam! And at Lock 12!
There was not even a *lock* at Lock 12. It was just a rapids-infested site between
two high bluffs surrounded by forest. The Corps of Engineers had once said
it was a good place for a lock if Washington (or Montgomery) ever wanted
to open the Coosa to navigation, but the Corps had since backed off the
project, so there was no lock. Most Chilton County folks figured there would
not be a dam either.[1]

Events that followed seemed to confirm this opinion. Though work started
at the site in 1909, it quickly came to a halt, and locals figured the whole
thing had been abandoned. But Lay kept buying up land, which gave rise to
rumors that the dam was just a ruse, that he was "talking that stuff to keep

us from knowing what he was really up to." William Patrick Lay, some concluded, was just "some crank who is hunting a gold mine out here in these hills." It had to be that. Anything more was simply beyond the imagination of most. "I'll tell you one thing," a Chilton County farmer reportedly told a friend, "I've been on this river a long time [and] nobody's going to put a dam across the Coosa." Even if they could, which the farmer doubted, no one could afford it. Why, he concluded, "if they did, it'd cost now on to $10,000."[2]

Locals had good reason to believe that even if Lay was telling the truth, the project had little chance for success. The Coosa was a wild river. Men had died trying to navigate her rocks and races, and after decades of plans, surveys, and some construction, the Corp of Engineers had concluded that the cost of taming rapids and shoals with names like "Devil's Race," the "Narrows," "Hell's Gap," "Butting Ram," and the "Devil's Staircase" was too great and therefore abandoned the effort. So the stream was left to fishermen and to an occasional scientist like Herbert H. Smith, who a few years earlier had collected snails and mollusks in the rapids and frequently "got a dowsing" for his efforts. But the river was not just big and dangerous; it was also isolated. Lay's point to Congress (if the dam broke, no one would be hurt) was only a slight exaggeration. There were few people in the area to help Lay build his dam, no roads to carry them and their equipment to the site, and no place to house them. The only things the Lock 12 site had to offer were high bluffs and a flowing river, but to William Patrick Lay, they were enough.[3]

Rolifee Bates was one of the skeptics. He and his family lived about a mile from the river, and some of the 1,500 acres he owned was on or near the stream. When Captain Lay was exploring the region, camping along the Coosa and looking for land to buy, he stumbled across the Bates home. Not wanting to disturb the household needlessly, Lay returned to Clanton where he made some inquiries and found that Bates owned the land he wanted. It followed that a short time later six-year-old Belle Bates looked up to see a "thin little man" get out of a buggy and walk up to her father, who was working in the field. Captain Lay had arrived.[4]

Things did not start well between the two men. Lay told Bates he wanted to buy land, but Bates did not want to sell. Then Lay told him about his plans, and Bates, who surely knew about the abandoned construction down-stream, "just laughed and told the captain he was crazy." But William Patrick Lay kept talking, and soon the conversation moved from the field to the porch. There they sat for a long time, while Lay related his "dream of a dam and electricity," and the farmer listened. Three quarters of a century later, in 1990, Bates's daughter, now Mrs. Belle Hendrix, remembered the scene and the talking. When they were done, she recalled, Lay had convinced the doubtful Bates that "he could do it . . . [for] he sold him the land."[5]

■ ■ ■

Memory and timing do not always fit neatly; recollections run together, and it is difficult to map out that clear chronology dear to the heart of historians. Therefore, it is hard to say just when Lay appeared at the Bates home, but it all seems to have taken place some months before Lay turned the Alabama Power Company over to Martin and Mitchell. What we do know is that with this change of command, the project got under way in earnest. Back in the spring of 1911, some men who were interested in "investing in Chilton County" went out to look at the work Lay had started earlier and came away unimpressed with the project and the progress. If they had returned six months later and seen the land being cleared and the construction crews at work, their attitude would have been very different indeed.[6]

With Congress's 1914 deadline facing them, James Mitchell and Thomas Martin moved quickly. They put the final touches on an agreement with the MacArthur Brothers Company of New York, a firm that arrangements reveal had both the experience and the expertise to handle the construction of the dam and powerhouse. Signed on August 1, 1912, the "cost plus fixed fee" contract put MacArthur in charge of the day-to-day management of the project but gave Alabama Power Company "unlimited powers in the direction of the work" and the right to "exercise close supervision over all its detail"— tasks that fell to Chief Engineer Eugene A. Yates and his assistant, E. L. Sayers, to Design Engineer Oscar G. Thurlow, to Electrical Engineer W. E. Mitchell, and especially to Resident Engineer A. C. Polk. Thus authority was divided between company and contractor, which seemed a good idea at the moment. By the time the job was finished, a lot of people were not so sure.[7]

The contract called for the project to be completed by December 31,

Contractor's office at Lock 12,
September 1912

(Alabama Power Company Archives)

1913, with a bonus if finished earlier. By August 5 survey parties were in the
field marking the elevations for the dam and the reservoir, and core samples
were being taken to be sure the bedrock would support the structure being
planned. Meanwhile, MacArthur Brothers began to bring in its people—a
superintendent, his assistant superintendents, an office manager, and a master mechanic (who happened to be kin to the superintendent, a relationship
later deemed "not in general a wise one"). All these were employed by the
contractor and directly responsible to the New York office. Though they were
expected to work closely with Alabama Power, and though the company had
some authority over them, their loyalty clearly lay elsewhere.[8]

17

Problems began to surface from the start. Even before work at the dam
began, MacArthur Brothers' choice for Railroad Superintendent was "re-
lieved from duty . . . for drinking," a loss that delayed work on the line being
built to carry goods the twenty-two miles from the Louisville & Nashville
railway station at Ocampo to the site. This, of course, delayed almost everything else. The next superintendent proved little better and was also let
go. The third and last "got material over the road in better shape" than his
predecessors, but according to Resident Engineer Polk, none of them was
"entirely satisfactory on this job." Although the contractor was able to over-

come, or at least work around, most difficulties, the railroad was a problem from the start of the project until its completion.[9]

While the contractor worked to get the railroad built and operating, prospecting parties went into the field to find sources of sand, gravel, and other material needed for concrete. The search was all but futile. Two quarry sites were located that could provide some of the stone needed, but no suitable sand or gravel was at hand. These critical items, along with much of the cement that was used, had to be brought from Montgomery, Leeds, and as far away as Richard City, Tennessee, which increased the cost of operation and made efficient transportation all the more important. Despite these early problems, the project went ahead with "very little friction" between the various departments. Indeed, Polk noted that "at the height of the work [there] seemed to be a spirit of co-operation and a desire to make high records" among the units. That was a good sign.[10]

Even though most of the railroad between camp and Ocampo followed the bed of an old lumber spur, forty-five trestles had to be rebuilt, so it took over sixty days to get the train running. In the meantime crews began clearing the land that would be flooded. Surveys showed that the Lock 12 reservoir would cover some 4,700 acres, and of these, 1,700 were "more or less heavily wooded." The rest were either in the stream itself, creek bottom farms, or land that had been cut over by lumber companies. Although a "thorough investigation" by the company had shown "that there would be no actual ill effects to the health of the community at large" if they left the timber standing, it had been advised that "the temperament of the hill people residing in the neighboring territory, and their respect for the law, in so far as it may assist them in obtaining compensation for imagined damage, made it appear prudent to clear the reservoir." Thus warned, Thomas Martin invited Dr. W. H. Sanders, state health officer, "to make an inspection of the proposed reservoir," at company expense, to be sure that the work would "meet the approval of . . . the health officers of the state." That visit also seemed prudent, and time would prove that it was.[11]

Local sawmills were contracted to cut timber and prepare it for use in the camp buildings as well as in the construction of the dam itself. Over 4.5 million feet of the 6.6 million feet used came from reservoir land. The rest

Clearing Lock 12 reservoir. The object was to cut and remove
the trees so there would be no danger to navigation and to burn the
branches, tops, and brush so there would be little trash to float
down to the dam and the powerhouse.

(Alabama Power Company Archives)

was bought nearby, but the price was cheap ($10 per 1,000 feet), and the
goodwill the purchase generated among the locals who preferred cash to
trees made the bargain even sweeter. Following health department suggestions, clearing crews were instructed to leave standing trees that would be
completely submerged when the reservoir was drawn down. The other trees
were ordered cut "as near the ground as can conveniently be done," the logs
removed, and the trash burned. Crews were also to cut ditches to drain pools
where water would be left standing when the lake was lowered. These were
simple directions, and the company was confident that they would be carried
out.[12]

Cofferdam at Lock 12, closed and unwatered. Cofferdams were built in sections (cribs) around an area of river bottom. When the cofferdam was complete, the water inside was pumped out, and the riverbed was exposed. At that point construction could begin.

(Alabama Power Company Archives)

By the end of 1912 work on the dam was beginning to attract attention beyond the neighboring counties. Reporters from Birmingham visited the site and came away "profoundly impressed" with the undertaking and with the men behind it. By January of 1913 construction on the first cofferdam was well advanced. This was a key element in any dam building operation. It involved constructing a temporary structure of rock- and dirt-filled cribs extending out from the west bank and into the river. When the cofferdam reached a point at or near midstream, it curved down and around to return to the shore. Then the area enclosed in the coffer was pumped dry ("unwatered" in engineer jargon) so that the foundation and first stage of the dam could be constructed. When they were built, the first cofferdam was torn down; while the river rushed through stream control openings in the just-completed portion of the dam, a second coffer was started from the opposite

21

side. This cofferdam was built much like the first and pumped dry so the next phase could be constructed. Finally, if all went according to plan, the two sections of the dam would be joined, the powerhouse completed, and with the river flowing through the culverts at the base, concrete would be poured to raise the structure to the required height. Then, one by one, flap-gates over the openings would be dropped into place, the culverts would be closed, and the reservoir would fill.[13]

On paper and in outline, dam building seemed simple enough, but in reality it was a complex process that required scores of skilled workmen and hundreds of laborers. Working around the clock in ten-hour shifts, laborers found the work often backbreaking and dangerous, as hospital records reveal. Construction required machines and methods the likes of which had never been seen by most residents of Chilton and Coosa counties. It also required financing at a level far greater than most of them could imagine. One newspaper tried to give its readers "a general idea of the magnitude of the work" by pointing out that "the money being spent [at the site] equals the total official building operations in the city of Birmingham during 1912." It was considerably more than the $10,000 the farmer believed it would cost to dam the Coosa.[14]

Dam building was not the sort of work men from Alabama farms and hamlets were accustomed to doing, but even if it had been, there were simply not enough of them to do the job. No sooner did the contractor start hiring than unemployment in the surrounding region disappeared, and the local labor pool dried up. Before long MacArthur Brothers did what one would expect of such a firm—they turned to New York labor agencies, who recruited "foreign laborers" to fill the breach. Brought from New York by sea to Savannah, Georgia, and from there by rail and road to the site, these immigrants made up the bulk of the "common laborers" first employed at the dam. For some reason, however, they "were not satisfactory" at that location, so the imports were moved to the quarry, where they "did so much better that for a long period the quarry labor camp was almost exclusively of foreigners." Soon, however, the Power Company concluded that "the loss was so great on the transportation of foreigners from New York that labor of this class was finally abandoned." Other sources would have to be found.[15]

At this point Alabama Power took labor recruiting into its own hands.

Black workers had been on the job from the start, doing the same common labor that the foreigners did. Observing this condition, company managers concluded that "the negro, with all his unreliableness, [was] the most satisfactory" solution to their problem and hired a "regular labor agent . . . who knew the Southern Labor Market well" to seek out recruits in places much closer than New York. The agent proved his worth, and soon a "constant stream of men" was arriving at camp. Florida was the main source of this supply, but Mississippi, Tennessee (especially Memphis), and Birmingham also contributed their share. Transportation was paid by the Power Company and deducted from the worker's wages. A problem even greater than recruitment proved to be retention, for the work was hard and dangerous, and living conditions left much to be desired. Therefore, the company promised to reimburse the worker if he remained on the job sixty days. This helped some but not enough. From the beginning of the job the labor situation was, according to an engineer's understated evaluation, "very troublesome."[16]

Skilled labor was a problem as well. Local whites were judged "absolutely unreliable, and of an exceedingly poor class," so they "did not afford any help whatever in solving the labor question." As a result, the contractor sought out journeymen workers who followed "big jobs of this kind around the country." Rather than seek "mechanics, hoist runners, machinists, etc. . . . from similar jobs scattered all over the South," MacArthur Brothers turned again to northern sources, and soon the Lock 12 site was a babble of different tongues and accents. Many of the carpenters they employed were Swedes from New York City, who "proved far more satisfactory than the local talent," and there was a large contingent of Italians as well. In fact, imported skilled labor proved generally reliable, and there were few if any complaints about their work. Like other journeymen who came to the project, they did their job, took their pay, and left when they were through.[17]

■■■

As construction moved along the size of the work force grew rapidly, and by the summer of 1913 the *Montgomery Advertiser* estimated that "no less than 1500" men were on the job. Despite its size, the operation seemed to be running like a well-oiled machine. From the superintendent at the main office down to the "flunkies" and water boys making ten to fifteen cents an hour, everyone appeared to perform the task allotted them. Night and day, at the dam, at the quarry, and on the railroad, they worked. By mid-December, with the west cofferdam completed and the riverbed pumped dry, round-the-clock work began on the first section of the dam. The day shift cleaned and prepared the forms; then at night, in the glow of "flaming arch lamps," concrete was poured. The next day the forms were removed, cleaned, and set in place for the night operation. Supervisors in charge of the various departments saw to their duties and passed orders down to foremen and walking bosses who oversaw the labor itself. As the dam slowly and steadily rose, the quarry turned out stone to be crushed for gravel, the railroad daily resupplied the site, and soon the second cofferdam snaked its way into the river from the east. All of this construction was accomplished under the watchful eye of Power Company engineers and two United States government inspectors, who were on the site to certify that "a proper foundation was secured and that suitable concrete was put in." By the end of the summer of 1913, things were going so smoothly that it seemed MacArthur Brothers might receive the bonus promised if they finished early.[18]

In some ways the order and progress was deceptive, for labor was "troublesome" not only in the recruiting but also in the regulation once the men were on the job. Anticipating this trouble, either from experience or from general assumptions about the nature of the workers they attracted, the labor agents and MacArthur's representatives put in place strict rules and routines that they expected workers to follow. When recruits arrived at either the dam or quarry camps, they were registered, given a number, and assigned their quarters. They took meals at the camp mess hall, and the cost was deducted from their pay, as was their rent. What money remained after this weekly settling-up could be saved, sent home, or spent at the commissary. Credit at the commissary was extended only to a worker who had been on the job long enough to pay off his transportation, a condition designed to keep men on the job but also an irritant that encouraged them to leave. When a worker finally served out his probation, he was given a "brass check" that allowed him to run up biweekly tabs at the store, and soon a black market in stolen checks flourished around the camp. Monthly credit was extended "only to responsible high class mechanics, and heads of departments,

etc.," whom officials believed would be on the job at the end of the month
to settle up. Other workers, those considered less dependable, had to pay
their bills each payday.[19]

Camps, a main one at the dam site and others at the two quarries, were
built to house the workers, and for a brief time the one at Lock 12 was not
only the largest but also the most cosmopolitan town between Birmingham
and Montgomery. The Lock 12 village was unlike other towns in the region
in ways other than size and ethnic diversity. In the first place, it was more
segregated. In small southern towns blacks and whites usually lived close together, separated by tradition and custom rather than physical space. At
Lock 12, as well as at the quarries, things were different. The main site was
actually divided into five well-defined sections. The largest was the white
camp, which contained the commissary, white mess hall, bunk houses, housing for the contractor's supervisors and their families, and most of the supply and storage buildings. Nearby was the engineers' camp, which provided
lodging and office space for Alabama Power employees and their families as
well as a "Guest House" for "the accommodation of Company guests and
visitors." Then there were a third camp for foreigners, mostly Italians, and a
fourth for the Swedish carpenters, who "lived off to themselves."[20]

Last was the "negro camp," about one-fifth of a mile from the main camp
and across the railroad. Worker housing in southern industrial villages in
Birmingham and elsewhere had long been strictly segregated by race, and
Alabama Power followed those examples. From the company's perspective
this separation had a number of advantages. First, the division showed ob-
servers that black workers would be kept in their place and that they would
not be treated in any way that would suggest social equality. Visitors took
special note of this and were assured by it. At the same time, supervisors
believed that blacks "had to be handled right" if their work and conduct
were to meet company standards, and "handling right" involved policies different from those applied to whites. Management believed that isolation
from the general population would make blacks easier to control. Isolation
would also lessen the chance of racial confrontation, which neither the company nor contractors wanted. On one hand segregation made it easier for
management to treat black laborers differently from other workers, while on
the other hand separation made it easier to protect them as well.[21]

Negro quarters at the Lock 12 camp

(Alabama Power Company Archives)

Blacks were not the only workers treated differently. The Italian and Swedish camps had no mess halls, because (the company explained) "this class of labor prefers to board themselves." The Italians were particular about their bread, and to keep them happy a special bakery was established to "bake the kind of bread espccially preferred by them"; at the quarry camp some were provided with "little kitchens" where they could "do their own cooking." Being practical men as well as welfare capitalists, the contractor and the company realized that "unless proper food and comfortable habitations were provided it would not be possible to hold the men," so they made a special effort to keep workers satisfied in this regard. Sanitary drinking water, "ice daily during the hottest months," fresh bread baked in the company bakery, and plenty to eat at reasonable prices were available to every worker, black and white, foreign or American, and though whites were given better jobs and treated with more respect than African American employees, blacks found that conditions in the camp were in many ways better than those they had left behind in a sharecropper's shack.[22]

Black workers were just as insistent as Italians and Swedes when it came to their needs and desires, and while records of Italian demands reinforce the often-cited Italian love of good food prepared as only Italians can, black demands, as recorded by white management, underscore the image of the fun-loving, free-wheeling Negro. Soon after the camps were occupied, black workers apparently came to their supervisors to ask for changes, and their requests were granted. Management's explanation was that since it is "essential to holding the negro laborer on the work that he have his negro women with him, as he will not stay in the camp unless the women are there," black women were to be allowed and "provisions [were] to be made to take care of them." It is not clear if the women were wives or girlfriends, and quite possibly the supervisors made no distinction between the two, but obviously black workers, seeing families in the white camp, wanted the same privileges—and got them.[23]

Though management probably would not have admitted it, blacks at Lock 12 were doing what blacks had always done, and done with remarkable success—they were confronting the system and forcing it to compromise. Sufficient labor was and would continue to be a problem, and black workers knew that their position was stronger than it might appear on the surface. Although they were careful not to make demands that would put them at odds with white supervisors and employees, at times they were able to get extraordinary concessions from the company and the contractor. The same report that told of black women in camp also revealed that management had found (or had been informed) that "another feature of holding the negro is that he be allowed to shoot 'craps' and play 'Skin,' his favorite pastime and card gambling game," and had concluded that it was "useless to try to stop him." Games of chance such as these, of course, were illegal, and the Chilton County sheriff's department was ready to see that the law was enforced. But a violator arrested was a worker off the job, so the company and county discussed the matter and decided that as long as "the negroes did all of their gambling and card playing in the confines of their own quarters," they would not be "molested there." The gamblers "were not allowed to go elsewhere and play, and Company police forces saw that order was maintained." Apparently white workers neither made such demands nor received similar concessions.[24]

This is not to suggest that the white camp was a model of order and self-control. Once off the job, whites were just as inclined as their black co-workers to conduct themselves in ways that confirmed management's assumption that men of their "class" needed regulation by their "betters" because they were unable to regulate themselves. But how they might be regulated once again reflected the common belief that blacks and whites were fundamentally different. After construction at Lock 12 was finished, recommendations for future projects included the suggestion that next time "a responsible white man who thoroughly understood the negroes" should be hired to act as "camp boss." It would be his job to keep "an eye on bad negroes," make sure black workers did not "lie around camp unless actually ill," and "in general maintain order" in that section of the village. In contrast, this postmortem called for less stringent measures in the white camp and suggested instead that "drinking and gambling" there could be reduced or eliminated if the company simply provided "a large, airy, reading and writing room" for the workers and "placed it under the charge of a competent man." There white workers could gather, engage in productive pursuits, and (as an extra advantage for the company) be in a place where "the superintendents can put their hands on some of the men quickly in times of emergency." Despite the different measures to deal with camp problems, it is apparent that, while the dam was being built, local bootleggers saw to the needs of both races, and officers of the law had more than they could handle.[25]

Still, one should not get the impression that the camps were some sort of a cross between a prison and a casino, which is far from the case. Situated on the west side of the river, high above the stream, the camps were well drained and in most regards attractive. A school was erected for the children in the white camp, mostly the sons and daughters of supervisory personnel, and on Sunday it doubled as a place "for religious gathering." In the black camp the company built a church, which apparently would have doubled as a school if there had been black children about. On this point the records are not clear, but it seems that while a few wives and sweethearts did live in the black camp (just as some must have lived with white laborers), traditional family life was most common among those workers whose occupations kept them on the job from start to finish. On later projects provisions for families were added as a way to keep workers at all levels happy and productive, but at Lock 12 the company was just beginning to learn that there was more to building a dam than pouring concrete and installing turbines.[26]

In addition to seeing that workers were well fed and adequately housed, the company and contractor took steps to see that the camps were sanitary so no epidemic diseases would disrupt the schedule. Attention to the health of workers and care for those who were hurt on the job was a matter of business as well as an humanitarian concern, and early on the MacArthur Brothers drew up a contract with Dr. P. I. Hopkins of Clanton to provide medical services for employees. MacArthur agreed to erect a clinic and temporary hospital of "ordinary construction" for the doctor and his assistant and to help Dr. Hopkins equip it. The clinic would treat mostly "out patients" who could recover, if need be, in their quarters. Hospital cases would be sent to Clanton. Dr. Hopkins agreed to reside in the camp, in quarters provided, as would his nurse and assistant. He would also "follow out the instructions" of the MacArthur superintendent, with two specific exceptions: he would not "attend women and children," and he would not "treat venereal and private cases." Under this agreement most workers would be assessed $1 a month, of which 90 percent would go to the doctor for his salary and expenses. At its peak the job employed over 1,000 workers and during the roughly eighteen months of the doctor's contract averaged from 400 to 500, so Dr. Hopkins did very well indeed.[27]

The doctor also served as sanitation officer. Under his direction a "pail system" of "fly proof closets" was installed to serve as toilets. The "refuse" in these facilities was "kept well sprinkled with lime"; every night it was collected, "hauled a considerable distance from camp, and buried." In addition, "all kitchens and dining-rooms were screened, as well as practically all the family dwelling-houses." Though later reports would chide the contractor and doctor for not doing more, inspections at the time reported that "the general health of all the camps was remarkably good." Except for several cases of typhoid fever at the Italian camp, which were blamed on "bad meat secured from an outside source," there were no outbreaks of "contagious or infectious diseases." On the whole, sickness had little impact on the workers' ability to get the job done.[28]

Accidents, however, were another matter entirely. Pictures of work at Lock 12, as well as at later sites, show scores of men working under conditions that would never have been allowed under modern rules and regulations. Heavy equipment, hundreds of hand tools, board and timbers, dynamite, rocks of all shapes and sizes, tons of concrete, a wet, slippery surface on which to work, and a roaring river always at hand created a host of safety hazards and countless opportunities for injury. The added fact that most of the common labor had never before worked on a job such as this resulted in a contractor's nightmare. Despite all the dangers that faced the workers, a final assessment of the project concluded that the number of accidents at Lock 12 was "not excessive, but on the contrary very reasonable when the nature of the work [was] considered." Everyone, worker and supervisor, seemed to accept the fact that "dam construction necessitates handling of many materials over the heads of workers and of their climbing and working in dangerous and risky places where it is difficult and impracticable to provide safeguards." Accidents that did occur were those "common to all large public works," so an employee who hired on was one who knew "of these risks and in general takes his chances and accepts them."[29]

Between the fall of 1912, when actual construction began, and November 1913, there were 382 accidents at the dam, in the quarries, and on the railroad. Most of these were classified as "non-preventable," and of the total, 95 percent were considered of a "minor character." There were six fatalities during the same period. Most of the injuries were treated on the site, and in many cases the worker returned immediately to his job. The company admitted that some of these mishaps might not have happened if there had been more rigid rules and regulations. However, the resident engineer pointed out that "an enormous amount of time would be lost and the job seriously delayed were such [a system] put into effect and strictly enforced," so nothing was done. Besides, the report rationalized, such a system probably would not work anyway, for the men generally ignored rules that stress safety

General view of the dam at Lock 12 under
construction, showing the foundation of the powerhouse
and the start of the form for the drafting tubes
that would guide the water through the turbines

(Alabama Power Company Archives)

over efficiency and instead took the "shortest and quickest routes" available
to them, regardless of the hazard. Therefore, if the job was getting done and
the workers accepted conditions as they were, the company concluded that
there was little incentive to do things otherwise.[30]

This opinion soon changed. Not long after construction began supervisors complained of "shyster lawyers" from as far away as Birmingham who
had set up offices just outside the confines of the camp and who, with the
help of paid representatives among the men, tried to get injured workers to
let them handle their cases. Promising "a large return in the way of damages
from the Power Company," the attorneys (according to company reports)
left "no stone unturned to get at all injured" and convince them to bring suit

before the company insuring Alabama Power could "make reasonable and quick settlements." These lawyers hired "women friends of injured men" to smuggle papers to them and urge them to sign; automobiles were sent to camp "after dark" to pick up workers and take them to hospitals where doctors friendly to the lawyers could assess their injuries; and some visitors, claiming to be relatives of the patient, proved to be in the employ of the attorneys. The company instructed officers at Lock 12 to arrest these people for trespassing. Despite increased efforts, however, they got in, and it was not long before there were twenty-one injury cases in court. Eleven of these, supervisors noted, were handled by the same firm.[31]

The insurance company, which should have been Alabama Power's ally in this situation, proved little help at all. Rather than acting quickly when an incident occurred, the insurance company dragged its feet for days, even weeks, which frustrated the injured and gave lawyers time to contact them. MacArthur Brothers took note of this delay and offered to share the cost of having an insurance agent on the scene to deal with cases as they happened, but the proposal was declined. Nearly a year passed between the time construction began and the arrival of an insurance company representative to inspect the site and point out "any special or ordinary safeguards" or call the supervisor's attention to anything that had been overlooked. The result of the visit was a short list of recommendations that were "immediately carried out," and that was it. On-the-job supervisors could conclude only that "the Insurance Company was either well satisfied with the condition of things, or very slack in protecting both their own and the Power Company's interests." Therefore, the lawyers continued to be active around the camp, and Alabama Power continued to go to court.[32]

Meanwhile, work on the dam went forward on schedule. The first cofferdam was sealed, secured, and unwatered by mid-December 1912, and work soon began on the first portion of the dam. This section went up through the spring of 1913; then in the summer, as the Coosa fell, cribs of the first cofferdam were taken down and water flowed through the stream control openings. On June 17 work began on the second, shorter cofferdam, and by summer's end construction on the second portion of the dam was underway. At the same time the tailrace, where the water would reenter the river after flowing through the powerhouse, was excavated. The powerhouse was also built during this time and made ready for the delivery of the turbines and other machinery that would generate the electricity. By December of 1913 the contractor's part of the job was all but done.[33]

Although MacArthur Brothers had failed to earn the early completion bonus, they had come in on schedule, no mean accomplishment. By the end of the year some of the small gates had been closed, and a pool was forming behind the dam. Everything was ready, and just before midnight on December 28, the first large flap was lowered over its opening. Then one after another, at five-minute intervals, the other gates were cut loose, and by 2:00 P.M., they were all in place. The seams were caulked, and workers began concreting the holes. Rain had been falling for the past few days, so upstream the water "rose fairly rapidly." On January 1, 1914, the Coosa flowed over the crest of the dam and back into its bed below. Inspectors looked for leaks around the flaps and along the foundation. They found none. The Coosa was dammed.[34]

All that remained was getting the powerhouse into operation. MacArthur Brothers' work was finished, and the company turned the project over to Alabama Power Company and its engineers. The bricking and ironwork were still to be done, and more concrete had to be poured, but these operations were minor compared to what had gone before. A 100-ton-crane was brought in to maneuver the turbines and generators into place, and during the next few months that massive machine went about its task with slow, steady efficiency. Electricians and mechanics followed in its wake, hooking up the lines and making everything ready. The first unit was put into service on April 12, 1914, and the other three soon followed. Now the Alabama Power Company could be what its name promised—a *power* company.[35]

By then most of the labor force had been paid off and sent home. All that remained in camp were a few skilled workers, a clean-up crew, and the supervisors and engineers. Power flowed across transmission lines that were erected while the dam was built, and everything worked as it should. No one would have faulted the men who made it happen if they had put their feet up on their desks, kicked off their shoes, lit up a cigar, and taken a well-deserved break. But they did not. Even as the project was finishing on a high note, some of the men closest to the work were preparing an evaluation of the project that candidly called attention to things that went wrong and

Flap gates ready to be closed at Lock 12.
When these were closed and sealed the
reservoir would begin filling.

urged that changes be made in the future. Although no one talked yet in specifics, there was no doubt that the leaders of Alabama Power did not consider this dam their last.[36]

A. C. Polk, the resident engineer, and E. L. Sayers, who served as the assistant chief engineer, were the most forthright in their assessments. Little escaped their notice, whether it was the shortcomings of the cableway system used to transport material across the river or the distance between houses in the Negro camp. However, two aspects of the operation—the commissary and the medical department—were singled out for special comment. Neither Polk nor Sayers liked the arrangement the company made with

MacArthur Brothers and with Dr. Hopkins. Guaranteed a fixed fee, the doctor apparently sought ways "to cut his expenses to a minimum," which was done "to the detriment of his service to the men." Polk and Sayers recommended that next time the company organize a medical department under the superintendent of construction, one that would "not [be] run for profit or gain, but purely and simply for the benefit of the men." The doctor and assistants employed would be "regular members of the organization" and therefore more loyal to the company than someone contracted from the outside would be. This doctor would also act as a sanitary officer with authority to enforce regulations he deemed appropriate—another weakness of the system set up by MacArthur Brothers.[37]

The engineers also found fault with the commissary department, though less in what it did than in what it did not do. On the whole the commissary worked well enough. Supplies from Montgomery and Birmingham arrived on time and were sufficient to meet the basic needs of the camp. The problems lay in the building, which was small, cramped, and remote from the men, and in the variety of goods, which was to say, limited. Polk and Sayers recommended that in the future "the commissary store be a large airy building, centrally located, [with] regular counters, shelves, and tables" on which would be displayed a variety of "good clothes, candy, [and] fancy groceries." The writers found it "astonishing" that laborers on the job would "spend all they make" on such luxuries and that when they could not get them at the commissary, they would flock to "a store operated by a Jew, just out of Alabama Power Company property." The owner of this establishment stocked these "attractive" items, charged "big prices" for his merchandise, and made "a good profit." The report concluded that if the commissary paid more attention to what the men wanted, as well as to what they needed, the workers would be happier and the company would have yet another source of income.[38]

The assessment by Polk and Sayers contained many other recommendations, including the size and location of houses in the camp, ventilation and drainage, sanitation, ice and water, and recreational facilities, but on the whole they congratulated MacArthur Brothers for the way they handled the details of construction and for getting the project finished on time. Nevertheless, they concluded that there were "certain features which in the light of experience gained could be improved on in a similar piece of work." What they did not say, but what was apparent in their report, was that many, if not most, of the problems might have been prevented if a contractor more directly responsible to the Power Company had been in control of operations. Divided authority might work well when all sides involved had the same goals and expectations, but when they did not, problems were sure to arise. The next time, and everyone associated with Alabama Power assumed there would be a next time, things should be done differently.[39]

Almost Done in by a Mosquito

When they first built th' dam, Ol' Man Mac come down with chills so hard that he shuk th' bed posts, an' he liked to died. It was a long spell 'fore he was able to be up, but when he got outa bed, th' first thing he done was to go hell bent fer Rockford. He got a lawyer thar an' sued th' power company fer th' biggest figger of money he could think of.

—WILLIE BASS, Coosa County

ONSTRUCTION OF THE LOCK 12 DAM WAS ALL BUT ignored by the local press. Newspapers in Birmingham and Montgomery gave the project special attention, perhaps because they were more aware of its statewide significance, but in Clanton, Verbena, and Rockford activities over on the river got no more space than the death of a prominent citizen and less than reports on cotton prices in Mobile. On one occasion the *Rockford Chronicle* did caution readers that "dam business . . . [was] not a vulgar expression, although some people have tried to use it as a 'cuss word,'" but there was little in the article to suggest that, apart from adding spice to the language, the project was having much of an impact. It was almost as if local editors did not quite know what to make of it all, and because the site was far removed from the towns and workers caused little trouble in the neighborhood, there really was not much "news" to tell. Readers in the big cities might be impressed and enlightened by articles on engineering innovations with pictures of the dam at various stages of construction; Birmingham and Montgomery might be interested to learn that a party headed by Gov. Emmet O'Neal had gone "by special car to inspect work on the Coosa"; but apparently the car did not stop in any Chilton County towns, so there was nothing to report. Besides, a wedding, a good recipe for bread-and-butter pickles, or a fight at a local roadhouse made much better reading.[1]

Then, late in the summer of 1913, reports from the basin above Lock 12 set tongues to wagging, and soon the local press began to take notice. Although company officials must have heard rumors of what was taking place down there, their first formal notification may have been the letter from Dr. H. L. Castleman, health officer of Talladega County, that arrived at the company's main office in Birmingham the first week in September. In it Castleman informed Martin, Mitchell, and their associates that "numerous complaints have been received from the lower end of Talladega County regarding conditions in connection with your development at and about Lock #12." To underscore the seriousness of the matter, the doctor added that the "complaints come from the most influential people of the locality & while they may be without just cause, possibly it were best that they not be ignored." Things had been going very well. The first section of the dam was finished, the second coffer was completed and unwatered, and now this occurred.[2]

Thomas Martin, as chief counsel for the company, had no intention of ignoring the doctor's warning, for although the complaints were not listed, they obviously related to the office that Castleman held. Even before work on the dam began, Alabama Power was concerned about health conditions in the camps and in surrounding communities. The company knew that in 1911 and 1912 nearly 150 suits had been filed against the Central of Georgia Power Company complaining that a dam it had built had created a stagnant reservoir full of drifting logs and decaying trash that caused "impure air that is odious offensive, and unpleasant, [and that] emits therefrom unhealthy odors, smells, vapors and gasses, malaria, [and] miasma." Thomas Martin feared that the same thing might happen to Alabama Power. Early on he was advised that the "hill people" living around Lock 12 were ready and willing to turn to the law to "assist them in obtaining compensation for imagined damages," and it was with that warning in mind that Martin invited state health officer Dr. W. H. Sanders, along with health officials from Coosa, Chilton, Shelby, and Talladega counties (including Dr. Castleman), to inspect the clearing that was taking place in the proposed reservoir and advise the company on what needed to be done. It was, Martin wrote in invitation, "the desire of the power company that the work in which it is engaged shall meet the approval of all the civil authorities, as well as the health officers of the state." It was also his and the company's desire that, if suits were brought, officials should be available to testify that Alabama Power had done what the state had asked it to do.[3]

The main thing that concerned Mitchell and Martin was how much time, effort, and money should be spent clearing the land that would become the lake. Wanting to do what was necessary, but not wanting to do more, the company engaged Prof. Edgar B. Kay of the University of Alabama as a consulting engineer to consider "the question of the inundation of timber lands in relation to public health" and advise them. The professor rose to the occasion, and his report, which obviously pleased Thomas Martin, was sent along to Dr. Sanders. In Kay's opinion, the removal of all vegetable matter from the basin before it filled was necessary only if the water was to be used for drinking, which of course it was not. Therefore the water might not look, smell, or even taste quite right, but it would be "perfectly wholesome" for other uses. To underscore his point, Kay noted that the federal government was apparently unconcerned with such matters, since at its Warrior River project "not a dollar has been appropriated for the purpose of timber removal." As for mosquitoes and malaria, Professor Kay observed that water in the reservoir would still be moving downstream and mosquitoes did not thrive in moving water. Moreover, he added with admirable optimism, the reservoir might just "improve the sanitary conditions along the banks . . . by submerging the swampy places and low grounds" where mosquitoes breed. All of these considerations led the good professor to a conclusion that Alabama Power wanted health officers to read and note. He could see "no reason from a sanitary viewpoint why lands should be deforested as a prerequisite to inundation or submergence." The company might want the timber to use in the dam, but any other reason for clearing the area that would be flooded was "purely sentimental." Local residents might protest and demand that Alabama Power go beyond what was necessary, but if the company went further the action would, according to Kay, "be a question of expediency not of public health."[4]

Professor Kay's letter was apparently in Sanders's hands when Martin and Chief Engineer Eugene A. Yates appeared before the state board of health in January 1913 to ask for a "recommendation or statement . . . as to what was necessary to be done" in the basin. With clear guidelines to follow, the company believed it could carry out the work quickly and at the lowest possible cost. More important, however, Martin knew that if cleaning crews followed recommendations from the board of health, the board would support the company if its work was challenged. But Alabama Power also wanted regulations that would be easy to follow, so in his presentation Yates reiterated most of Kay's points and put special emphasis on the opinion that the reservoir was "not what in the ordinary sense of the word is meant by a lake . . . [for] instead of the water remaining quiet or being stagnant . . . it is always running." Therefore, Yates concluded, there would be no mosquito problem, so "the only point to be considered" was whether trees that were submerged would "decay and produce odors injurious to the health of the people living in the immediate neighborhood." The board, Yates argued, should not be too concerned on this point either. "If no clearing whatever is done," he continued, "no sickness of any nature will result . . . due to the odors," which will only be "noticable . . . at the spillway where the hydrogen sulphide is liberated." "Continual movement of the water will prevent stagnation" and keep down mosquitoes; "trees partially submerged will die and fall and in time decay"; and those "fully submerged" will die and be preserved underwater "for an indefinite period." "The only condition which might cause sickness," Yates asserted, would be if stagnant ponds where mosquitoes could breed were left when the reservoir was drawn down in the summer. This, he concluded, would happen whether the land was cleared or not and could be prevented "by proper draining." Like Professor Kay, Yates saw little reason to clear much of the basin and even less cause to be concerned for the public's health.[5]

Still, Yates wanted the basin cleared, for if it were not, he suggested to Martin, "innumerable law suits will be brought against the Company." On this point at least Professor Kay had been right; "it was a question of expediency not of public health." E. A. Yates was a realist. He believed there would still be suits, even if the company spent the estimated $60,000 cost to "clear the reservoir entirely." But if the land was cleared, Alabama Power would have a better chance of winning in court than if it were not. Therefore, even though Yates knew such an undertaking would "set a precedent" for future projects, he believed that it would "be in the end less troublesome and less expensive to clear the reservoir entirely, removing all trees to within

Driftwood deposits and stagnant water similar to these were said to create a health problem for people living near the Lock 12 Reservoir.
(Alabama Power Company Archives)

two feet of the ground and cutting and burning all brush." Martin, who had been in touch with lawyers for Georgia Power and had heard the same advice from them, did not need more to convince him. Instructions went out that the reservoir be cleared and what was not used be burned. State health officials approved this approach, adding only that ditches should be cut to drain the "standing pools of water" that would collect when the reservoir was lowered.[6]

When Dr. Castleman's letter arrived, word went out to supervisors at the site to send someone into the basin and see if health department and company guidelines were being followed. The results of the inquiry were not what Martin wanted to hear. The investigator reported that in various parts

42 of the reservoir trees and brush had fallen into creeks where they were rotting, stinking, and backing up water into stagnant pools where mosquitoes bred. All this created "a very ugly looking situation" and one that Alabama Power would be hard-pressed to explain. The company also learned that petitions were being circulated asking the board of health to take action and that feelings were running so high that it seemed likely that people in the neighborhood would "continue to have complaints and trouble" until the creeks were cleaned out. Aware that Alabama Power's management wanted to project a positive image to the public, the investigator noted that these conditions also made "a very bad showing" for visitors, for one of the creeks ran along the railroad "and passengers can and do, see and remark about the present condition" of the land. Lawsuits and negative publicity were two things the fledgling Alabama Power Company did not need.[7]

Back in Birmingham things were heating up. As reports came in Martin wrote Yates that it was "highly important . . . for this subject to be given close attention at once," and when Yates returned to the office Martin wanted to "discuss the matter personally." Meanwhile, a Talladega lawyer working for the company sent a letter containing more information on the "considerable agitation" occurring in the southern end of the county. He included with it a letter from one of the "prominent residents" of that region claiming that the power company was fouling streams with timber "which will create sickness, for it has already begun to sour." "We people," the writer demanded, "want the timber out and burned to preserve the health of the community." This was not just one man voicing his opinion, for a few days later Martin received a letter from Assistant Chief Engineer Edward Sayers telling of how over a hundred people gathered at a "Meeting of Protest at Talladega Springs" to denounce the company. But Sayers's news was good news. The meeting had been called by county health officers to inform angry locals that officials in Montgomery were convinced that the "Alabama Power Company was not only following the instructions of the State Board of Health, but [was] doing more." That said, the officials added that if anyone in the crowd decided to sue, "the State Board of Health would have to rise to the defense of the Alabama Power Company." "The meeting," Sayers reported, "was an entire success from our point of view."[8]

There was one small hitch in the proceedings. County officials added that if those doing the clearing "left any pools of stagnant water" the company "would then be guilty of offense and liable." That exception left an open door, and it was not long before the company was in court, brought there by residents claiming that stagnant pools were being created by trees cut into streams and not removed. The plaintiffs argued that these conditions "will greatly impair the health of the community," and as a result a number of the residents were "going to leave the neighborhood." Company lawyers discovered, much to their distress, that "witnesses who have heretofore been friendly" were among those preparing to depart. Although state officials took the stand on the company's behalf, without locals to testify in its favor, Alabama Power lost. It was only one case, but it sent a signal to the rest of the community.[9]

The jury awarded the plaintiff $550, a sum that caused considerable concern at Alabama Power headquarters. Because of the threat of war in Europe, Mitchell and Martin were anxious about capital resources and about the company's ability to pay outstanding debts. By the fall of 1913, the dam was nearing completion, but at no place in the reservoir was the clearing and burning finished. Water backed up by trash in creeks was problem enough, but if the gates of the dam were closed before the basin was clean, scores of suits would surely follow. Now orders went out for cutting and burning crews to double their efforts, and Yates assured Martin that when the cleaning was done Alabama Power would have "a reservoir that will be absolutely cleared of trees and brush with nothing projecting above the surface of the pond." "In other words," he emphasized, "we should have a perfectly cleared sheet of water."[10]

Company lawyers on the scene thought this approach would be satisfactory, as did state health officials, though one of them noted sarcastically that if timbermen had cut the trees "out upon the land instead of in the creeks they could have managed the situation with a great deal more ease and much less expense." But agreeing what needed to be done and doing it were two different things. Though county officials seemed convinced that "the Company was doing the work as rapidly as they could," it was not fast enough for locals whose streams were clogged with timber that had soured and whose water was "badly discolored" and from which "offensive odors arise." Fortunately for the company the weather was getting colder, and so the "eddy pools" around the logs and trash were not yet breeding places for mosquitoes. But locals feared for the future. Some became so concerned that, instead of going to court, they approached Power Company representatives and "implored" them "to buy the balance of their land . . . because they felt that it would not be safe for them and [their] families to stay near the pool."[11]

But for each resident who wanted to sell out and leave, there were many others who wanted to sue. Reports back to headquarters told how the "strong sentiment against the Company" that already existed was being "encouraged and magnified by parties with selfish motives." Martin was informed that at least one attorney had opened an office in Columbiana "for the expressed purpose of bringing damage suits against the Alabama Power Company," and the "common street talk" was that soon Alabama Power would be in court again. Some residents, those schooled in populist notions that had been standard fare in Alabama politics for decades, even claimed that "the Power Company [was] taking all this property for its own personal gain and that the people . . . [would] bear the greater burden and take all the risk of damage to property and health." Hardscrabble farmers and small-town merchants also took as gospel the rumor that the company would not serve towns such as theirs or people like themselves because they were "too small and cannot use enough power." Soon politicians got into the act, and a candidate for senator began soliciting votes on the promise that he "was for the people owning the water powers of the State of Alabama." All of these things—lawyers, politicians, "antagonistic sentiments," and wild rumors—were, according to company representatives on the scene, reasons "why we should remove every possible ground for criticism within our power."[12]

Company leaders understood that a problem of public health might soon become a crisis in public confidence. Reports coming back to Birmingham told of a widely held "feeling that we have not kept faith with the citizens in and adjacent to the Lock 12 Basin," and of how "they seem to be losing faith in the Company." From the outset public image and public support were important elements in the Power Company's plan, and now those elements were being threatened. The money was one thing, for the company could ill afford to pay damages in the cases that were being considered, but the loss of public confidence posed an even greater threat to the future. If Alabama

Power Company could not be trusted to use the public waterways for the public good, federal licenses would be denied, favorable legislation would not be passed in Montgomery, and sources of capital would dry up. In other words, without the trust of the public there would soon be no Alabama Power Company.[13]

Of particular concern to Thomas Martin was the attitude of health officials in Montgomery. Although they had supported Alabama Power in its efforts, and even testified on its behalf in court, as protests mounted and petitions were circulated, Dr. Sanders pressed the company to move faster and do more. However, the company, which had put extra crews into the basin after the first complaints, seemed to think things were going well enough. This impression was confirmed early in the new year when it sent an inspector to check on conditions at the reservoir. He reported the "situation very satisfactory as a whole" and noted that none of the local landowners he interviewed "seemed to be fearful of trouble." But not every inspector found the same conditions. Shortly afterwards Dr. Sanders sent his own representative, a man he considered "both competent and reliable," to visit the site. This inspector saw and reported timber floating in the lagoons and "not a single ditch cut" to drain the water from the pools that formed when the lake was down. Dr. Sanders received the report and forwarded it to the headquarters of Alabama Power.[14]

When he received Sanders's letter Martin concluded that it was time for Alabama's chief health officer to see things for himself. He immediately contacted the doctor to assure him that it was "the purpose of the Company to comply with the conditions heretofore laid down by the State Committee of Public Health" and to suggest that if he would make a "personal inspection of the reservoir . . . we can come to a clear understanding as to what if anything should be done." Sanders accepted the offer. About a week later Alabama Power representatives met him in Clanton and took him to the dam. There they boarded the company boat for the tour. Apparently it was a complete success. Dr. Sanders returned to Montgomery, and there were no more letters critical of the company's work in the basin. Soon spring came. The reservoir was full, the creeks and swamps were flooded, and the air was clear and sweet. There were no "unhealthy odors, smells, vapors and gasses." The crisis seemed to have passed.[15]

Then during the summer of 1914 trouble appeared again—this time in the company's own backyard. At Lock 12, the former construction camp was now a village where employees who operated the dam lived with their families. It should have been one of the healthiest places in the region. Houses there were screened, a permanent sewer system was being installed, residents were inoculated against typhoid fever, and those who showed signs of malaria were dosed with quinine. Still, people got sick, and malaria was the biggest problem. During the summer of 1914 it was reported that "nearly all [the] families . . . stationed at Lock #12 have had some member or members ill with a serious attack" of the disease. By fall the matter had become of such concern that the company sought out an expert on tropical diseases to advise them.[16]

Of course malaria had been prevalent in the Coosa Valley as long as locals could remember. In the summer months mosquitoes always swarmed out of the creek bottoms and swampy margins along the river, infesting homes and businesses, annoying both man and beast. Even without the streams there would have been a problem. Every farm in the region had animal troughs, rain barrels, and trash dumps with cans and bottles full of water, ready and waiting for mosquitoes to lay their eggs. So common were these conditions that some families were known for the sickness among their members, and as long as there were such people to act as hosts, the disease was sure to spread. The health problem was so severe that some officials suggested that malaria had cut cotton production almost as much as the boll weevil had. In rural Alabama, where cotton was king, this matter was not to be taken lightly.[17]

In the summer of 1914, however, sickness throughout the region seemed worse than ever, and locals did not take long to conclude that the culprit was the Lock 12 reservoir. Soon the company was "receiving complaints daily" from people who claimed that because of operations at the dam, the level of the lake sometimes dropped as much as four feet, leaving "a number of pools of water throughout the reservoir, which became the breeding places for mosquitoes." What followed surely came as no surprise. Late in September four suits were filed in Shelby County claiming that this fluxation of the water left "large areas of ground covered with dead fish, tadpoles, frogs, and insects," not to mention "brush, stumps, bark, leaves, muck, slime, and other

Lock 12 Dam and Reservoir. Note the extent of the area flooded and the inlets and sloughs where driftwood could collect.

(Alabama Power Company Archives)

sediments and filth." This, it was alleged, became "sour, foul, and in a decaying condition," and "the water became infested with mosquitoes, tadpoles, and insects, plasmodum and malaria." If that was not bad enough, plaintiffs also claimed that the water in their wells had become unfit to drink, their homes were "infested with mosquitoes," and they were all "annoyed by unpleasant odors." Put simply, they claimed that their land had become "undesirable and sickly" because of the Power Company's negligence and that they had been "compelled to abandon the property." For the losses they suffered,

48 for their medical expenses, and for the "mental pain and anguish" they had endured, each plaintiff asked the sum of $1,400—nearly three times the earlier judgment that the company had paid. And lawyer Martin understood that those suing did not have to prove every allegation to win their case. This matter was serious indeed.[18]

Martin knew by reports from Lock 12 that malaria was a serious problem around the reservoir, and he reasoned rightly that these suits were only the first of many. But the cases would not come to trial for nearly six months, which gave the company time to investigate conditions and prepare its defense. In this interval he could gather those "officials, doctors, and sanitary experts" who had advised Alabama Power in the past, convince them of the "soundness and defensibility of their earlier reports and conclusions," and keep them "in a cooperative frame of mind" until they were called on to testify. No one at corporate headquarters doubted the importance of this preparation. Not only would the decisions reached in these cases determine whether or not other suits would follow, but also it was understood that "as the issue is determined here so we may expect to see it determined in connection with the Cherokee reservoir when that shall be built." So with an eye on the immediate *and* the distant future, Thomas Martin prepared his case.[19]

Things did not look good for Alabama Power. Chilton County's health officer investigated the charges and reported to Sanders that "thousands of logs [were] found floating" in the reservoir and that "thousands of trees, some dead, some dying, and others still living [were] partially submerged." In his investigation he "saw no ditches for draining pools" and observed that "the water at some places has a very offensive odor." Worst yet, sickness was everywhere, and in one case a teacher "abandoned his school before it was out because of his being confined to his bed with Malaria." Hoping for a better report, the company requested that the health officer make a second visit, and accompanied by Alabama Power representatives he toured the reservoir again. His conclusions were much the same as before. Although he praised the company for creating "one of the largest and most beautiful bodies of water in the South," he noted that when the party left what was once the riverbed and turned up the small tributaries and sloughs that led into the lake, things were "not pleasant to sight or smell." Moreover, among the natives the inspector found "scores having chills where none had malaria in former years, and many moving away, some leaving their cotton unpicked." "True," he admitted, "the people—some of them—are very backward and live very crudely," but (and here he underlined to emphasize his point) "*it is their homes,* and health conditions are driving them away."[20]

Other health department investigations found similar situations, and to make matters worse, when one of the company's own, Design Engineer O. G. Thurlow, toured the lake he returned to report that "conditions about the head waters of the reservoir . . . [were] contrary to my expectations and not at all satisfactory." Logs and brush lying about had hampered drainage, and in some low places the land had "taken on the nature of a swamp." Unless something was done, Thurlow warned, things were "bound to become worse." The area, he admitted, would probably have been swampy even if it had been "thoroughly cleared," but that likelihood was beside the point. "In this locality," Thurlow concluded, "we have not strictly abided by the requirements of the state health officials." This statement was not what company managers in Birmingham wanted to hear. They knew that if the conditions Thurlow found could be shown to have caused the sickness that plagued the area, the Alabama Power Company was in trouble.[21]

Locals had no doubts as to the cause or the responsibility. Willie Bass, a sharecropper, bootlegger, and fisherman from Coosa County, believed, as so many did, that the "fever come when th' water was backed up over th' trees. . . . they beginned rottin' an' that put th' fever in th' air." His landlord, "Ol' Man Mac[,] come down with chills so hard that he shuk th' bed posts, an' he liked to died." When Mac recovered he knew what to do. He went "hell bent fer Rockford," where "he got a lawyer," and "sued th' power company fer th' biggest figger of money he could think of." Ol' Man Mac was not the only one. Just as the company feared, as the year drew to a close, attorneys in the counties around the reservoir were doing a brisk business as residents came to them with their complaints. Optimism among the litigants ran high, and many, like Mac, thought "the money was as good as [theirs] already, an' there wasn't nothin' left to do but cash th' check." Local merchants extended credit on the expectation of large settlements and, for the moment at least, their business was as brisk as that of the lawyers. Alabama Power Company representatives had promised locals that the dam

Alabama Power Company boat *Mapleleaf*
inspecting Lock 12 Reservoir. The logs
floating near the bank were said to create
breeding places for mosquitoes.

(Alabama Power Company Archives)

would bring prosperity to the region; no one anticipated that lawsuits would
be the way it was brought.[22]

Meanwhile the Power Company began to gather support for its conten-
tion that the reservoir was not to blame for the sickness in the region. Another group of county health officers spent two days "inspecting the sanitary
conditions of the lake, . . . and the tributary streams . . . and backwaters,"
and although they encountered rotting logs, trees, and brush just as others
had, their conclusions were markedly different. These officials went one step

further and investigated sanitary conditions around farms and homes where
malaria occurred but which, in many cases, were far from the reservoir.
There they found mosquito larvae in stagnant water near barns, at wash
places, in hog pens and cow pens, in old wells, and in branches. They also
learned that many residents were having chills and fever years before the
dam was built. As for the lake itself, there they found only "one or two larvae
in a little ravine touching the river," and those might well have washed down
from above. All this evidence led the team to conclude that "local conditions
for germinating mosquitoes, should and must be removed before attributing
the prevailing chills and fever to the reservoir or lake."[23]

As 1914 became 1915 the tide seemed to be turning in the company's
favor. Surveys revealed more malaria carriers in locations well away from the
reservoir, reinforcing earlier conclusions that Alabama Power's impound-
ment was not solely to blame for sickness in the area. Meanwhile Martin had
found health officers and other witnesses who would testify "that the conditions of the reservoir had nothing to do with the malaria of the plaintiff and
other members of his family." The company was almost ready to go to trial.
Only one thing was missing. Back in September 1914, when the first suits
were filed and Martin began to plan his defense, it was suggested that in
addition to a parade of friendly witnesses, Alabama Power needed someone
who could not only testify to the facts but who could also "be impressive
before a jury." Such a person, however, had not been found, and it appeared
the company might go to trial without one. Then someone thought of Dr.
William Crawford Gorgas—Surgeon General of the United States Army, the
man who defeated malaria and yellow fever in Panama, a recognized authority on the habits of mosquitoes, and a native of Alabama.[24]

Martin and his team of lawyers may have been thinking of Gorgas for
some time. Earlier in October a member of the state board of health had
written the General for information on mosquitoes and the spread of malaria, and a copy of that letter, along with Gorgas's reply, found its way into
the Alabama Power Company files. No doubt Martin had read the letters and
was aware that Gorgas doubted that a reservoir such as the one at Lock 12
would be the source of sickness in the adjacent counties. Martin and Gorgas
probably exchanged correspondence on the matter before Martin went to
Washington to ask the surgeon general to come to Alabama, visit the lake,

investigate the charges against the company, and give his opinion on "the whole matter." "Whatever experience I have," Gorgas responded, "should be available for the well-being of the people of my native state and of the entire South." So the man that Thomas Martin hoped would "impress the jury" agreed to make the trip. But Martin surely knew it was a gamble, for if Gorgas found that the company was indeed at fault, every case against it would be lost. By February 1915 more than 700 damage suits had been filed. A lot was riding on this visit by William Crawford Gorgas.[25]

Engineer Thurlow was chosen to escort the general, who arrived as scheduled, and together they "covered all of the territory on both sides of the river where [there were] suits pending and investigated both the conditions around the houses and conditions in the reservoir." There had not been much rain in the past few weeks, so the breeding places for mosquitoes were much as they had been the previous summer, and Gorgas could get an idea of what conditions would be during the height of malaria season. For three days they made their inspection, and when they finished the general informed the company that he would testify on its behalf.[26]

The case being heard was, according to Thurlow, "one of the worst suits which we had to contend with," so a victory would set an important precedent. Knowing this significance, Alabama Power's legal department put an extraordinary effort into its preparation and even went to the trouble to take the jury down to the reservoir to see things for themselves. In the end their hard work was of little consequence, for, as Thomas Martin later commented, "William Crawford Gorgas was enough." Gorgas, the "unimpeachable authority" and an Alabama hero in the bargain, took the stand and (again according to Martin) "gently, clearly, simply in language all could understand, as if he were conducting a class . . . explained how short-lived and short-ranging [the malaria carrying] mosquito is, . . . [and] that it is unable to fly more than a short distance from its breeding place, certainly not as far as the complainants' homes from the lake waters." The general then told of "his visits and examinations around the vicinities of these homes . . . and of this or that pool or ditch of stagnant water" he discovered nearby. "These, and not the reservoir, . . . were the cause of the malaria." The jury, whose members had visited the same sites and seen the same things, agreed. The twelve men retired and "in less than thirty minutes" returned to announce that the Alabama Power Company was not at fault.[27]

William Crawford Gorgas
when Surgeon General

(From Marie D. Gorgas and Burton J. Hendrick,
William Crawford Gorgas: His Life and Work
[New York, 1924], 312.)

One can only imagine the joy and relief felt by company officials. Not only had they won, but they knew, as the *Birmingham News* noted in its article on the verdict, "that this case will go far towards disposing of the various other cases pending in other courts." According to the *News,* Martin and his team of lawyers had addressed the charges "from every standpoint," and it appeared to the reporter covering the case that "nothing further can be shown by any other of the parties who have sued the company in such matters." Meanwhile, down in Columbiana, and in Rockford, Clanton, and Sylacauga, locals and lawyers began to reassess their situation. The tide had

54   turned against them, and one after another they decided to withdraw. Things returned to normal. Another spring was coming, and it would soon be time to plant.[28]

A few of the more hopeful, or more desperate, did pursue their cases, but to no avail. Willie Bass remembered how "Ol' Man Mac . . . went to curt lookin' like he was already dead, an' he had a chill or two sittin' right up thar . . . but th' jury didn't pity 'im." But it was not the jury that Bass blamed for the outcome. Revealing old populist prejudices that would keep some people suspicious of Alabama Power and its motives to this day, he concluded that Mac "might a knowed that they ain't no pore man can win nothin' off'n th' power company." Years later, when Willie Bass made this pronouncement, the Great Depression was at its worst. Unemployed, he was living on company land near the Coosa, land that Alabama Power rented to folks like him for $3 a year, which it never collected. Still, Bass believed that those nameless, faceless people who ran things in Birmingham "look at us folks hyar on th' river like we'uns was no better'n a dog." Against the Power Company, he concluded, "a pore man don' stand no more chanct than a June bug in January."[29]

Willie Bass—sometime fisherman, sometime bootlegger, and full-time river rat—was not typical of the residents who lived in the counties that surrounded the lake. Hardworking farmers, close-trading merchants, and the usual collection of small-town professional men and women continued living as they always had and looked forward to the day when the power being generated at Lock 12 and sent to cities like Birmingham and Montgomery would light their homes and businesses as well. There was a war in Europe now, and reports from France and Russia filled the local press. Cotton prices were up and promised to go higher. Alabama Power, its sources of credit restricted by conditions overseas, rearranged its financial structure so it could seek capital from American investors. This change prevented the company from undertaking a second project, so instead it set about to make Lock 12 a model of efficiency and expertise. Meanwhile, down at the dam and up at the home office, engineers and executives evaluated what had been accomplished and made plans for the future.[30]

Gathering Streams
from Waste

*. . . to gather the streams from waste and to draw from them energy, labor
without brains, and so to save man-kind from toil that can be spared, is to
supply what, next to intellect, is the very foundation of all our achievements
and all our welfare. . . .*

—OLIVER WENDELL HOLMES, *Mount Vernon–Woodberry Cotton Duck
Co. v. Alabama Interstate Power Company* (1916)

LTHOUGH THE ALABAMA POWER COMPANY HAD
been absolved of responsibility in the "mosquito suits," everyone
knew that there was malaria in the land, and company employees were just as susceptible to the disease as was the population
in general. In one of his frequent assessments of conditions on the Coosa,
Design Engineer Oscar G. Thurlow acknowledged that residences near the
reservoir were "typical for the great part of the state," and he believed that
"the lack of progress among our farmers can be attributed more to malaria
than to anything else." Thurlow's investigations revealed that many of the
people who sued the company had "been having malaria for years," and in
his judgment they went to court because the disease left them without "am-
bition enough to get out and hustle for a living by farming or some other
means." Where earlier studies claimed the hookworm was responsible for
Southern laziness, engineer Thurlow placed the blame squarely on the mosquito.[1]

Not wanting to lose time and men to the disease, Alabama Power sought
advice from the U.S. Public Health Service on measures that would reduce
malaria at Lock 12, and then in the spring of 1915 the company invited officials to come to the site and see what had been done. Apparently Thurlow,
who escorted the group, expected the visitors to lavish praise on the village,
but to his surprise the engineer found himself "in a somewhat embarrassing

position" of having to explain why "many of the most important preventive measures had been overlooked." To the casual visitor the camp looked "very good indeed," but closer inspection found "two operators sick in bed with malaria" and another who had "had three chills during the past week." It did not take them long to discover why the men were ill. There were "a good many potential breeding places for mosquitoes" on the grounds, and the screening that was supposed to keep the insects out of the houses had been so poorly installed that it was declared "an absolute waste." Even the superintendent's residence, which was pointed to as "the best screened house in the camp," was "not properly screened at all." To make matters worse, the "closet" toilets around the camp were "not in good condition as far as protection from flies go," and Thurlow was warned that the camp was "in a very good position to have an epidemic from typhoid fever." Before the inspectors left they handed the superintendent, J. U. Benzinger, another list of preventive measures. This time the recommendations were carried out.[2]

Soon houses and community buildings were rescreened, in some cases double-screened, and breeding places in the village were discovered and drained. Toilets were improved, and work continued on the permanent sewer system for the whole village. In addition, blood samples were taken from residents, and those with malaria in their systems were dosed with quinine to remove the parasites and prevent them from becoming carriers of the disease. Getting rid of malaria among the employees and their families would do little good, though, if outside the reservation there were people among whom "chills were so common [they] no longer took any notice of them." Rather than have these folk reinfecting residents of the village, the company decided to make quinine treatments available to all, and for a while Alabama Power ran something of a public health service in the neighborhood. Some residents refused to take the foul-tasting medicine, but others accepted the offer and for the first time in years found themselves free of the disease. For one price the company purchased both good health and goodwill.[3]

Like so much that was learned in building the dam at Lock 12, efforts to create a healthy environment for residents in the region (and keep the company out of court) taught Alabama Power the importance of preventing the conditions that gave rise to sickness and complaints. To control the mosquito population the company applied "oily waste" to possible breeding places, including the sloughs and inlets in the reservoir. This solution was not satisfactory, however, because the oil had to be applied over and over again, and it left a messy, smelly residue. So when Alabama heard of the *Gambusia* minnow, which was said to eat mosquito larvae, naturally it was interested. Here was a solution that, after the expense of stocking the lake, required no further outlay and no upkeep. Quickly company officials contacted the U.S. Public Health Service in New Orleans and the Bureau of Entomology in Washington for advice on how to set up a hatchery for the small fish. The *Gambusia* seemed to be the answer to the mosquito problem.[4]

Insects were not the only thing that created concerns for the Power Company after the dam was built. Local men had fished the Coosa for years, some for food, some for fun, most for both. Those who fished for the pot and the skillet often resorted to less-than-sporting measures to fill their larders, and among these dynamite was a particular favorite. What worked in the river also worked in the lake, and soon word of explosions and dead fish reached county commissioners. Hoping to put an end to the illegal practice, officials asked the company to recommend "some active, vigorous, vigilant man who resides in the vicinity of the lock" to serve as "deputy game and fish warden" and enforce conservation statutes at the reservoir. Locals were not excited about having another warden to carry out laws they did not like, and one of them suggested that the fish were "being killed by electricity more than by dynamite." This theory reached Thomas Martin who, amused at the idea, responded that it was a "physical impossibility" for "electric power carried by the lines of this company" to have been involved. However, he did admit (tongue firmly in cheek) that there was always the chance that "a wire should be broken and actually fall on the river bed and strike the fish," but that was unlikely.[5]

Meanwhile, Martin and Mitchell were still trying to clear the way to begin work at Cherokee Bluffs on the Tallapoosa River, the project they had wanted to undertake from the start. There was still opposition, especially from the Mount Vernon–Woodberry Cotton Duck Company, a factory powered by a 100-year-old water wheel. Mount Vernon–Woodberry's management believed that a dam upstream at the bluffs would restrict its water sup-

Mount Vernon–Woodberry Dam,
Tallapoosa River

(Alabama Power Company Archives)

ply and reduce its production, so the company went to court to stop the project. Lower courts upheld Alabama Power's right to build and sustained Martin's argument that the Cotton Duck Company would not be injured, but the plaintiff appealed each decision, so finally, in 1916 the case reached the Supreme Court of the United States. In arguments filled with populist rhetoric (a case of the "Water Power Trust Against the People of Alabama") and biblical illustrations (Alabama Power was out to steal the Tallapoosa as Ahab stole Naboth's vineyard), attorneys for Mount Vernon–Woodberry made their case. Thomas Martin responded in kind. Then the High Court ruled that Alabama Power was not taking Mount Vernon–Woodberry's property without due process of law and that the Cherokee Bluffs project could proceed. Justice Oliver Wendell Holmes, who wrote the decision, handed down the Court's ringing endorsement of hydroelectric power development: "to gather the streams from waste and to draw from them energy, labor without brains, and so to save mankind from toil that it can be spared, is to supply what, next to intellect, is the very foundation of all our achievements and all our welfare." Martin later wrote that "few cases in the annals of water power transcend the importance of this litigation."[6]

The next year, 1917, America went to war, and for the rest of the decade Alabama Power naturally devoted much of its resources to that national effort, especially in connection with providing electricity to nitrate works in the Tennessee Valley. In one case, however, the war came close to home. Down at Lock 12 a man with a thick accent was seen "hanging around the power plant for some days." Law enforcement officials were notified, and the man was arrested as a German spy. Though he claimed he was Austrian, the distinction did not matter to folks in Clanton, who held him until federal authorities arrived to take him away.[7]

For Alabama Power the war years witnessed a change in leadership that was to have a profound effect on the company's future. By that time it was obvious to most in management that Thomas Martin had become the driving force in the organization. In part this change was due to Martin's own ability, but contributing as well was the declining health of James Mitchell, the company's president. Mitchell grew less and less able to carry on his duties, and finally he knew the time had come to step down. Late in 1919 the board of directors, to no one's surprise, decided Martin should succeed him. Early the next year the change became official. Mitchell became chairman of the board, and Thomas Martin was elected president. Five months later, on July 22, 1920, James Mitchell died.[8]

Mitchell had wanted Alabama Power to develop an interconnected system of hydroelectric projects to serve the entire state. Martin took this vision one step further, and early on as president he signed an agreement with Georgia Railway, Light & Power Company for the exchange of surplus power between states, thus paving the way for a regional system even greater than the one James Mitchell had envisioned. But at the signing Alabama Power

60 was still a rising company, with only one dam in operation, Lock 12, and a steam plant on the Warrior River in Walker County, so just how much surplus power it would have was problematic at best. Now the war was over, and it was time to begin again. Negotiations with Mount Vernon–Woodberry were progressing but not concluded, so the company decided that a more practical project would be a second dam on the Coosa, at a place called Duncan's Riffle, about fourteen miles downstream from Lock 12. There Alabama Power would use the experience gained earlier to build an even larger hydroelectric facility and name it in memory and honor of James Mitchell.[9]

■ ■ ■

One of the earliest lessons learned at Lock 12 was that the company needed a closer working relationship with the contractor and that this might be best accomplished if the outfit building the dam was in some way a part of Alabama Power and thus loyal to it. Under such an arrangement Alabama Power would have its own construction force, trained in the work it was to do and familiar with local conditions where that work would be done. The men in this organization, especially the skilled workers, could move from job to job as needed, and for Alabama Power, with so many projects on the drawing board, that was no small advantage. In addition, a construction company such as this would be Alabama based and would therefore purchase most of its material and supplies within the state. This point was important, for the amount of money pumped into the local economy had a lot to do with the degree of support the Power Company received. Reviewing the problems experienced in working with MacArthur Brothers convinced Martin and his associates that they should move ahead with a new arrangement. In the fall of 1917 Alabama Power Company organized Dixie Construction Company and transferred the construction department to the new organization "in block."[10]

"The function of Dixie Construction," a 1925 company publication explained, "is to build dams and install generating equipment that produces electric power." Dixie did not design the project. Instead, Alabama Power's engineering and operating department drew up job specifications, produced the necessary plans, and prepared a cost estimate, all of which were submitted to Martin and the Power Company board for approval. Then everything was turned over to Dixie's construction manager, and work began. The engineering department inspected and supervised as needed, and when the job was completed the dam and powerhouse were transferred to the operating department. On paper at least the organization was all well planned and closely coordinated; duties and responsibilities were neatly divided and clearly assigned, and in the middle of it all was what Martin hoped would be the solution to many of their earlier problems—Dixie Construction Company.[11]

Everything was in place, and in November 1920, newspapers throughout the state announced that Alabama Power was seeking federal permission to build a hydroelectric dam at Duncan's Riffle. The press's attention this time was in marked contrast to the almost nonchalant attitude the media had when it told of the decision to build at Lock 12. People now knew the impact such projects had on the local economy. They had heard of how families like the Bateses in Chilton County had turned their home into a boarding house and how they had built three additional bedrooms and a porch to accommodate the overflow from the camp. They had also heard how Mrs. Bates ordered "groceries by the case" from Clanton merchants and how her "wood stove never got cold." For two years residents in the region shared the bounty brought by the Alabama Power Company, and so naturally they joyfully greeted the news that another dam would be built.[12]

Chilton County in particular was excited. At the beginning of the year Clanton had finally installed electric lights, an accomplishment, local boosters noted with pride, that put the town in the same league as Selma, Montgomery, and Birmingham. The event also gave the Wyatt brothers, Mack and Gene, the editors of *Clanton Union Banner,* an opportunity to turn their caustic pens loose on those citizens whose major occupation seemed to be "waiting for better times." As the editors saw things, with Clanton's homes and businesses lit by electricity, such citizens might discover prosperity "thrust upon them" whether they were ready or not. A "building boom" might follow, which would add new homes and "a number of fine store and office buildings" to the scene. Then, "for all we know somebody will come along and make us provide a location for a cotton factory or something." If Clanton was not careful, it was "bound to come to the front" in spite of itself. But it had not happened yet, so until success finally caught up with the commu-

Duncan's Riffle, Coosa River.
Survey work and clearing had just begun.

(Alabama Power Company Archives)

nity, the editors suggested that residents should be out there "fully enjoying the great advantages of that day we so long have sought when we might have as much juice as we wanted," and hope in the meantime that they can "get forgiveness for all the cussing we have done" in the past. By summer headlines in the paper announced that "Chilton County is Proud of Alabama Power Co.," and articles advised readers that company stock was "a splendid thing in which to invest your money." No one talked about mosquitoes and suits now.[13]

Locals were even more excited in November when they learned that Alabama Power would spend the grand sum of $8 million on the second dam. With everyone hurting from the postwar recession, people saw this as heaven sent, and the *Union Banner* suggested that "the farmers of Chilton County might find it to their advantage to arrange to work with the company in their construction work and make more money than they can hope to make in the next two or three years out of cotton." With the project here at hand, the editors noted that residents "can be the first on the ground with their applications and get the work if they want it." An "enterprise of [this] magnitude . . . [can] employ every farmer and laboring man in Chilton County," and, the article concluded, "we would be glad to see our people . . . get their part of this eight million dollars." The *Union Banner* had ruffled some local feathers in the past and would ruffle some in the future, but this time it was preaching to the choir. Folks were filling out those applications even as the editors wrote.[14]

Within days of the announcement, engineer parties arrived at the river and began surveying to determine the limits of the reservoir, as well as a route for transmission lines that would run from Lock 12 downstream to Duncan's Riffle. Operators at the dam "became so excited over the new development . . . that a bunch of [them] drove down [the next] Sunday to look over the site." Meanwhile teams evaluated possible routes for a railroad between the river and the L&N main line and selected Cooper as the connecting point, much to the disappointment of Verbena, which hoped to be the terminus. Then in January, without warning, work was suspended. The reason given was the "certainty of bad weather . . . and uncertain business conditions," but at the time the federal government still had not approved the project, and some believed that Martin did not want to spend more time and money on the effort until he had assurances from Washington. However, the suspension set off rumors in the region that the company might shift its operations over to Cherokee Bluffs on the Tallapoosa or abandon the project altogether, so folks in Chilton and Coosa counties waited the winter anxiously.[15]

But Alabama Power was committed to the Coosa, so in June 1921, when word arrived that the permit had been granted, work quickly began again. Initial surveys had long since been made, and land agents had already bought up much of the 5,000 acres that would be flooded when the water

64   rose. Those purchases represented the first real economic impact the project had on the community, and company leaders understood that on these transactions rested Alabama Power's reputation as a firm that would deal honestly with local people. Stories were circulated back then, and some are still circulated today, of how company representatives descended on unsuspecting residents, told them tales of foul odors and mosquitoes, and got them to give up their homes and farms at bargain prices. In some cases the stories may be true, but saving a few dollars on a few acres was hardly worth the damage it did to the good name of the firm and its future prospects in the region. Martin was on record opposing such tactics. The story goes that once, when the company was buying the right-of-way for a power line, the president was told by "one of the boys" that the land could be had for "$5 an acre."

"How much is it worth?" Martin asked.

"About $25," was the reply.

"Then we should pay $25," was Martin's response. "Pay what it's worth, but not a cent more. We must be fair, if we expect others to be fair with us."

By the time the final purchase was made above Duncan's Riffle, the company had spent $150,000—an average of $30 an acre. Not everyone was happy with the terms, and in a few cases condemnation proceedings were initiated, but on the whole, as the *Montgomery Journal* noted, by "paying satisfactory prices for the lands," Alabama Power was able to settle with "nearly all the individuals who owned property in the reservoir." That was good for everyone involved.[16]

Once the land was purchased, the clearing began. Clearing had been one of the most controversial aspects of the Lock 12 project, and Dixie Construction was well aware of the challenge it faced. Committed to helping the local economy whenever possible, officials on the scene suggested that the work "be done by men and teams now idle in [the] vicinity." But the workers and their mules were not simply hired and sent into the basin. Again, because of what happened earlier, the men hired were explained "the object to be attained" and carefully instructed to cut and burn all logs, branches, and brush that would foul places where "there was little or no wave action," taking special pains clearing land that would be "covered by a relatively shallow sheet of water." According to these guidelines, "no pine trees [should be] left

Sawmill located in the area being cleared
for the Mitchell Dam Reservoir

standing in water or close to water's edge" since "pine needles collected in
bunches" and were great places for the larvae. Had this been done before
there would likely have been few if any "mosquito suits," and Alabama Power
was not about to make the same mistake twice.[17]

As it was at Lock 12, much of the timber cut from the reservoir was used
in building the dam and the camp, but with the river above the riffle relatively free from rocks and shoals, it was possible here to float the timber
down in rafts to the mills. Floating the logs, "the majority of which [were]
a good grade of long leaf timber," was considerably cheaper than dragging
them by mule out of the forest, then shipping them by railroad to their des-

66   tination. So the basin was divided into milling units, with two large sawmills
and two smaller ones at strategic locations where the rafts could be broken
up and pulled from the river. When the logs arrived at these sites, they were
sawed and planed if necessary. Then the finished lumber was carried by
barge to the dam where it was unloaded. Before the project was completed
it had consumed over 5 million feet of timber, all of which came from land
in or near the basin.[18]

■ ■ ■

In the summer of 1921, just before it was officially learned that the federal government was issuing Alabama Power its Coosa River permit, a group
of men including one of the editors of the *Union Banner* set out for "that
God-forsaken place" known as Duncan's Riffle. Apparently Alabama Power
had not been very forthcoming about plans for the project, and since they
had heard rumors of work starting up again, they went there "to satisfy
[their] curiosity as to what was being done toward the new dam." The story
of the trip, published shortly after they returned, was a tale in the best tradition of Old Southwest humor, full of overstatement and satire, written
broadly for the biggest laugh, but containing within it insightful observations about what they saw and did not see. It was just the sort of thing the
Wyatt brothers had been writing for some time, and just the sort of thing
their readers loved.[19]

The visitors soon discovered that "the biggest job we had was getting
there and getting away," for once they were on the site "it was no trouble to
see what was going on while there." And what was going on? Not much.
There were "some indications that a dam, or something, may be built there
some day," but most of the activity the editor saw seemed to involve "a bunch
of surveyors" who had been able "to find the place by the aid of many square
yards of blue prints and a battery of spirit levels." On one side of the river
a "bunch of men [were] building pigpens on every rock that sticks its head
above the water of the river," and as soon as they were finished "another crew
fills them with rocks to keep them from drifting away." Curious about this,
the visitors approached some workers, but they "did not pay any more attention to their guests than if no one out of the ordinary were upon the scene."
They went "right on drilling or building pig-pens, or doing something else

and [did] not offer any information at all as to what their plans [were]." "Left entirely alone to form their own ideas," the editor and his friends concluded that though there were "some signs of something building down at Duncan's Riffle," they could not honestly say "whether it is going to be a dam or something else."[20]

The article was filled with the sort of jabs and jests that usually gave the boys down at the feed store or over at the barber shop a chuckle or two, and since locals were well aware of the editors' inclination toward this sort of humor, they probably laughed a bit and forgot the whole thing. But there were new folks in town, and to them what was going on out at Duncan's Riffle was serious business. A few weeks later a second article appeared, sort of an apology, that was written because it had been " 'hinted' to [the editor] by a friend that some of the local employees of the Alabama Power Co., did not like the tone of the article about the editor's visit . . . [and were] inclined to take it as expressive of some degree of animosity in the paper toward the Power Company." Normally if someone had been pricked a bit by the *Union Banner*'s editorial pen, the newspaper's response, if it responded at all, would have been "get over it." But in this case the Wyatts wanted to make it clear that those "looking for hostility to the Alabama Power Company" in the pages of the *Union Banner* were "simply on the wrong trail." "There ain't nothing doing in that line at all," they assured readers. "We are for the Alabama Power Company good and strong."[21]

The Wyatts went on to explain how the article was actually complimentary to the company, and the fact that none of the men on the job took a break to talk to the visitors was indicative of their attention to the task at hand. Though it was clear that the newspaper thought that Power Company folks who complained about the article were taking themselves too seriously, the editors wanted them to know that the *Union Banner,* in its own way, took them seriously, too. "Clanton is worth twice as much as a place to live in and do business in since the Alabama Power Company extended their system here than it was before they came," the writer told them. Furthermore, the editor and "the other gentlemen who went to Duncan's Riffle . . . [were] not mad at the way [they] were treated." No one in the party had any "ill feeling" over the matter and they hoped "no offense has been done to anyone on the side of the Ala. Power Co." But the editors could not resist one last dig. Had

Alabama Power been more forthcoming with information about its activities, there would have been no trip to Duncan's Riffle. Local folks needed to know what was going on. So, the editor concluded, "if the Company wants publication made of their plans and arrangements at Duncan's Riffle we shall be glad to know who can furnish us with the inside dope. . . . It would make a crackling good story and this editor would be glad to have the honor of conveying it to the public first."[22]

The Alabama Power was always of two minds when dealing with the press. On one hand management wanted newspapers to take notice of the company's achievements, for good press coverage could rally public support and attract investors. But company executives were also wary of what newspapermen might write and publish. Accounts of the "mosquito suits" and the Mount Vernon–Woodberry trial, articles outlining political attacks on company policies, and critical editorials exposed raw nerves, and in some instances Martin ran advertisements in key newspapers to explain Alabama Power's position on controversial questions and to counter negative publicity. No one should have been surprised, therefore, that supervisors out at Duncan's Riffle were less than receptive when the editor and his friends appeared and began "snooping around." But Alabama Power Company also knew that the *Union Banner* had a point, *the public* had a right to know what was happening on *the public's* river. All the company wanted was for the public to see things as the company saw them—was that too much to ask?

This attitude was surely one of the reasons why, in April of 1920, Alabama Power began publishing *Powergrams,* a slick monthly magazine full of articles and illustrations that was sent out to regular employees, stockholders, potential investors, and, of course, to newspapers, which reprinted pieces from it to update their readers on what was happening at various sites around the state. Lead stories were well written, carefully edited accounts that highlighted the organization's significant accomplishments. Filled with facts and statistics that were near and dear to the hearts of professional men from engineers to editors, *Powergrams* was a public relations success from the start. But the magazine was more than self-promotion and propaganda. Each department and each project had a column, and from the "news" sent in by someone at Lock 12, for example, readers could learn that children there

The second *Powergrams*, May 1920,
with Lay Dam on the cover

(Alabama Power Company Archives)

VOL. I.   MAY, 1920   NO. 2

Lock 12 Hydro Electric Plant of Alabama Power Company

PUBLISHED MONTHLY BY THE

ALABAMA POWER COMPANY

BIRMINGHAM, ALA.

70   were going to school just as children should be; that a new water system had been installed so residents "will have plenty of H$_2$O, minus the red mud, which the Coosa so generously brings to us all the way from Georgia"; and that "the only trouble" facing the supervisor was "getting a house ready for those employed." A local writer let folks know when spring arrived and the "hills [were] blue with violets" but also warned them that so many visitors had been coming to see the dam and enjoy the lake that the roads were in need of repair. These were personal pieces, almost like a Christmas "newsletter" from an old friend, and like those occasional updates, they were based on the belief that everyone knew everyone else and wanted to keep up with what they were doing. Collectively the articles left the impression that the employees of Alabama Power were one big happy family, which, quite frankly, was what Thomas Martin wanted the company to be. Today, folks long retired look back on those early days, and many recall that a family was just what Alabama Power was.[23]

Lock 12, the dam and the lake, were the pride of Alabama Power Company, and the editor of *Powergrams* delighted in touting its features and benefits. But downstream at Duncan's Riffle there was also plenty of which the company could boast. The September 1921 issue featured the Mitchell Dam site on the cover (the name was official now), and the lead article brought readers up to date on what had been done. At the "white village, a mess hall, construction headquarters, five engineer workmen's barracks and the temporary substation [were] completed and a large portion of the colored village was going up." Meanwhile, "outside the reservation" merchants had come from near and far to set up "stores ranging from a one-chair barber shop and bootblack stand to a two-story department store," and they had "changed this ancient cornfield into a bustling village." "Bustling" maybe, but "Dam Town," "Little Italy," or "Scrougeout" (three of the names by which this collection of businesses was known) was described by the *Montgomery Advertiser* as a "mushroom village of temporary shacks and a mixed population of cosmopolitan aspect." Where the editor of *Powergrams* saw this activity in a positive light and hinted that it might even become a permanent community, the *Advertiser* was struck by the similarity between activity there and the "mining camps of the boom days in the gold fields." Still, both publications agreed on what brought the people there. Everyone wanted their share of Alabama Power Company money.[24]

Back in the town of Verbena local citizens were feeling the effect of the company on their economy. When work first started, before the camp was built, many employees lived at the Porter House Hotel and spent their money in the town; but now there was housing at the site, the Porter House had taken on "a rather deserted look," and it was rumored that "since the engineers left the squirrels have been taking possession of the second floor." Verbena's city fathers realized that people were going to be paying attention to their area in the next few years, so they wanted to be sure that their town benefited from this exposure. Therefore local businessmen hired an agent to write an article touting Verbena's advantages and inviting individuals and businesses to come in and invest. Published in both the *Birmingham News* and the *Montgomery Advertiser,* the piece described how "the village, once popular as a Summer resort and now the trade center of a prosperous farming area, is building new business houses, making plans for industrial activities, and preparing again to blossom forth as a resort," all because of the activity out at Duncan's Riffle. It was boosterism at its best, and each businessman who underwrote the project had his interests highlighted. In it Alabama Power's impact was evident—from the newly opened bank to the "Dam Cafe." Investors from Birmingham had taken over "the once renowned Clifton Hotel" and were making changes that would turn it into "the most commodious road house on the Birmingham-Montgomery Highway," and the Star Store of Wetumpka ("the store that is different from them all") had opened a branch in Verbena. The town had all the amenities of modern life, including "Automobile doctors," and it invited everyone to come and share its good fortune."[25]

■ ■ ■

By November things were moving along well at Mitchell Dam. Transportation facilities were complete (including the narrow-gauge railroad that carried men and material around the site), and the "pig-pens" described by the *Union Banner* were well on their way to becoming the first cofferdam. The quarry was open, rock crushers were making gravel, and workers were getting ready to mix and pour concrete. Duncan's Riffle hummed with activ-

Camp at Mitchell Dam. Community
buildings (dining hall, etc.) are on the
left; worker housing is on the right.

(Alabama Power Company Archives)

Visitors inspect Lake Mitchell
from the company "yacht."

(Alabama Power Company Archives)

ity, and from every indication Dixie Construction was making the project a
model of efficiency. Alabama Power officials, recognizing how important a
well-run operation was to the company's public image, invited businessmen
and community leaders (including editors and reporters) from Montgomery,
Birmingham, and other cities to visit the site as the company's guests. The
natural beauty of the surroundings, which the *Birmingham Age-Herald* described as "wild . . . as in the mountains of Jefferson or Walker," added to the
attraction, and so it is hardly surprising that through the rest of 1921 and
into 1922 a "constant stream of visitors" came in "almost daily" to see what
was being done.[26]

Thomas Martin and guests at Mitchell
Dam, 1923. *Left to right:* Harry C. Abell
(Martin's brother-in-law), R. A. Mitchell,
S. Z. Mitchell, Thomas W. Martin.

Some arrived individually, others in small groups, but more often than
not they came in large parties, organized and coordinated by the company.
Soon after construction was underway, Alabama Power began sending representatives to speak to civic associations in various cities, and if they expressed an interest, these organizations would be invited to the site. In one
instance "about forty newspaper and businessmen from Anniston, Gadsden,
Sylacauga, Talladega, and Piedmont" took the tour. It began at Lock 12,
where they inspected the facility and enjoyed "a splendid luncheon." Then
the guests were carried by automobile to Mitchell Dam "where they were
able to see a modern power plant under construction." Late in the afternoon the party took the train home, duly "impressed" by all they had seen,
just as they were supposed to be.[27]

At times the company was able to provide visitors a special treat, like a quarry blast, but it seems that some among them were just as impressed by the "aerial trolley, 1500 feet long and operated over 100 feet in the air," that carried men and material across the river. They were also interested in "the great amount of construction equipment which [was] electrically operated," something company guides pointed out to emphasize, if it needed emphasizing, the versatility of the product Alabama Power produced. At times the guide was none other than Thomas Martin himself, who took special delight in showing people around and having his picture taken with them. Judging from photographs in the Alabama Power Company Archives, Martin was everywhere, and some who knew him recall that if a picture was being taken, he would find a way to be in it. According to their accounts, "he didn't believe a picture should be made without him."[28]

Of all Martin's trips to the site, the one he surely remembered best and treasured most took place on December 19, 1921, when he and a "throng of visitors," including some of the state's leading citizens, arrived at Duncan's Riffle to dedicate the project as Mitchell Dam. James Mitchell had been Martin's colleague and his mentor. At Mitchell's urging, and with his gentle guidance, Thomas Martin had been transformed from a promising young attorney to one of the most important industrial leaders in the state, if not in the region, and Martin knew how much he and Alabama Power owed to that hydroelectric pioneer. Indeed, it might have been more fitting if the dam at Lock 12 had been named for Mitchell, for that was his project. The structure that would someday rise out of the rocks at Duncan's Riffle really belonged to Thomas Martin. And since it was his, Martin gave it to his departed friend, and watched with pride and satisfaction as C. Malcolm Mitchell, James Mitchell's son, pressed the "electric button" that "tipped . . . the first bucket of concrete [that] was poured into the new dam."[29]

Building Mitchell Dam

*Business men and farmers in this section of the country are hopeful of
future conditions here due to the money that will be turned loose for labor by
the power company. It will come at the most opportune time in our history
too, for the dry weather has ruined crops in the eastern part of Chilton
County. There will not be any corn made and the farmers will welcome any
opportunity for outside work that will enable them to make a living.*

—*Clanton Union Banner,* July 14, 1921

G IVEN THE COMPANY'S EXPERIENCE AT LOCK
12, it is hardly surprising that so many things were done differently at Mitchell Dam. From beginning to end the job was better organized, transportation was more efficient, schedules were
more closely kept, the camp was better laid out and operated, and health
care for the workers was a matter of special pride. Lock 12 had been the
company's school, and in most cases students from Alabama Power learned
their lessons well. Each project completed during the 1920s had its special features, and the men who labored on them faced problems all their
own, but in each case Mitchell Dam was the example to which the company
looked when determining goals to set and steps to take.

It is also not surprising that the Mitchell project is much better documented than was work on Lock 12. Everything, including record keeping,
was centralized now, for the company understood that one day they might
need to refer to past activities. For example, health and hospital accounts
were carefully kept and preserved for the benefit of future planners and in
the event of more "mosquito suits." The company's increased interest in public relations also helped preserve Mitchell's story. Starting with its September
1921 issue, the one with the dam site on the cover, *Powergrams* included articles almost monthly that described in detail the work being done. In addition, newspapers published a steady stream of articles about the Alabama

77

Power Company and its activities. Many of these press pieces were prepared by company publicists and were as much propaganda as news, but even so they tell us a lot about what was going on out at Duncan's Riffle. They also tell us a lot about Alabama Power Company. Company leaders, from Thomas Martin down, realized that this sort of publicity produced not only goodwill among the general public but also money from investors, so with a sense of pride in what they were doing and the recognition that as managers of a "public" utility they could not ignore the public that sustained and supported them, corporate executives churned out the information. Once the dam was underway, editors like the Wyatt brothers in Clanton seldom had to complain that they and their readers did not know what was going on out on the Coosa.[1]

"Practically everything . . . [was] in readiness" when government approval was handed down, so "little time was lost in getting underway." The land had been bought, timber was being cut, sawmills were up and running, and lumber was arriving at the site where they would build the camp. Special attention was paid to the railroad, for company officials remembered "the difficulties that arose at Lock 12 because of poor railroad facilities" and knew "how much depend[ed] on having the best transportation system possible." The first question was where the line from the site would join the L&N. Verbena wanted to be the terminus, but that route would require more excavation and would take longer to put into operation; so the company chose Cooper, which was a little farther away but much easier to reach, and Verbena had to be satisfied with a dirt road to the river. Then there was talk of making it an electrified road, a nice touch for a power company, but since the tracks would be torn up after the dam was built, that idea was dropped. Work on the eight-and-one-half-mile line began on July 6, 1921, and ninety days later the first locomotive steamed to the site of Mitchell Dam.[2]

Even before work on the railroad began, even before they could begin improving the "highway" out from Verbena, Dixie Construction had to hire hundreds of workers, skilled and unskilled, to do the many jobs that had to be done. Where the labor recruiters of MacArthur Brothers had sought workers outside the South, Dixie turned to sources closer to home—with remarkable success. Even the weather seemed to cooperate in the effort, for that summer a drought "ruined crops in the eastern part of Chilton County,"

Mitchell Dam under construction.
A camp is located in the lower
section of the picture.

(Alabama Power Company Archives)

and according to the *Union Banner* farmers welcomed "any opportunity for outside work to make a living." News that Dixie and Alabama Power were hiring reached towns and cities throughout the state, and in Bessemer a group of enterprising businessmen bought three buses and established a passenger line between Clanton, Cooper, and the dam to carry workers to the site. Alabamians had seen the opportunities that came with Lock 12 and other Alabama Power projects and were ready to take advantage of what Mitchell Dam had to offer.[3]

Carpenters, among the first recruits, were immediately put to work on the camp and supply facilities. They were good, or at least fast, for by September 1, fifty-five buildings were ready for occupation and others were in various stages of completion. Once again Lock 12 served as an example of what not to do, for the camp at Mitchell was well planned from the start and offered workers amenities lacking on the earlier project. Divided into three parts—the white camp, the Negro camp, and a camp made up of family residences—it was located on wooded ridges separated by deep ravines. The white camp, with its bunkhouses and bathing facilities, was on the ridge nearest the river. It also contained the construction offices, many of the warehouses, the commissary, and the mess hall. On a higher ridge just above were eventually built twenty-four "comfortable cottages for white families," some of which would later form part of the permanent camp. Of course all the buildings had electric lights.[4]

As might be expected, the camp for black workers was carefully separated from the other camps. Anything otherwise was simply unacceptable to white Alabamians, so it is not surprising that both the *Birmingham News* and the *Montgomery Advertiser* made a special point of assuring readers that strict segregation was the rule at Mitchell Dam. The steep ridges and deep ravines that provided excellent drainage and helped improve health conditions also served as effective barriers, and though the two races lived only 400 yards apart, each camp was "entirely hidden from [the] view" of the other. Once again reflecting racial attitudes of the era, Dixie Construction believed it was far more important to control African American workers than their white counterparts. Eventually it decided that the natural barriers were not enough, and a ten-foot wire fence was built around the black village, supposedly "to prevent trespassers entering the camp." Though all the gates were never locked at the same time, Dixie located the entrances "at points under easy observation." White supervisors may have seen this plan as little more than a labor control device that was needed if order was to be maintained, but black workers must have been reminded of the prison road camps that dotted the Alabama countryside, an uncomfortable similarity that may have contributed to the high turnover rate among laborers.[5]

Inside the camp, however, Dixie Construction provided black workers' facilities that were far better than most would have found on the outside. Bunkhouses and bathhouses were much like those in the white camp, and blacks had their own mess hall. This was yet another indication of how rigid segregationist attitudes had become in the past decade, for now blacks and whites did not eat, even separately, under the same roof. As Lock 12 had taught them, planners made arrangements for black families with "20 two-family houses and one hundred and eighty 10 × 14-ft. shacks." Equipped with a "double bed, spring and mattress, and cast-iron laundry heater," the "shacks" (that was the term *Powergrams* used) were so popular that there was always a waiting list, and the company declared them "one of the best investments in the camp facilities." Since it was thought that "in order to hold large forces of negro labor it [was] necessary to provide amusement [as well as] quarters for the families," Dixie also built "a billiard hall and a dance hall" for the camp. But if an arrangement was made that would allow black workers to gamble in private as they did at Lock 12, that agreement did not find its way into the records. Just as in the white camp, bath houses and "sanitary toilets" were provided, and the bunkhouses and larger buildings had electricity. A "filtered water hydrant" was strategically located to serve groups of houses, but with the exception of the mess hall and the recreational buildings, there was no indoor plumbing. Conditions were not bad, but anyone living in the black section could see that in the white camps, things were better.[6]

Of course, conditions varied in the white camps as well, and Geraldine Hollis, whose father, John Hollis, eventually became superintendent at Mitchell, remembers how they called their section of the layout "poverty ridge" because the houses nearer the river were so much finer. Moreover, some whites, perhaps envying the individual "shacks" provided for black workers, made arrangements with the company to buy building material at cost and "erect small frame houses or tent houses for individual use." Some of these structures, later known as "U-build" houses, were even provided electric lights and water. But these social and racial divisions and distinctions reflected the times, and Dixie Construction's directors could not or would not have altered things if it occurred to them to do so—which it probably did not. Even though everyone acknowledged that building separate camps on separate ridges "greatly increased the cost of road construction, water supply, and maintenance," there was, at the time, simply no alternative. That was the way things were, and would be for years to come.[7]

The camps were laid out and appointed for the sake of efficiency, health, safety, and, it was frequently noted, beauty. Planners made sure that all portions of the premises were "well lighted and numerous roads and paths

82 [were] built to provide quick access to any portion of the work." All the buildings were "well separated for fire protection," and a filtered and chlorinated water system served the residents. Those parts of the reservation that were to be part of the permanent village were "provided with sanitary sewers," and the rest used a "dry closet system" (outhouses) where cans were changed frequently, and (at least as sanitation was concerned) "excellent results were secured." Situated as they were, high above the Coosa, with a commanding view "of the sweeping bend of the river, with its heavily wooded banks," the camps became noted for their natural beauty that as the families moved in was said to have been enhanced by "the presence of potted flowers and plants" in windows and on porches, which gave "evidence that the hand of woman [had] lent its refining influence there." With pride the October 1921 *Powergrams* reported that "shortly the camp[s] will be enjoying most of the conveniences of the city, with none of the congestion."[8]

As work moved forward it was clear that the company's public relation efforts were paying dividends. The *Birmingham News* published numerous articles praising the working conditions at Mitchell and the impact the project was having on neighboring communities. Not to be outdone, in early October the *Birmingham Age-Herald* sent a reporter to the scene. Having seen the mining camps in and around Birmingham, he was surprised to find at Mitchell a "modern construction village where sanitation, orderliness and comfort are chief considerations." Down on the Coosa, the *Age-Herald* announced, "the mill hovel with filth in front and more filth in rear has been replaced by a cozy home with bath tub and sanitary closet, with flower gardens and screened doors." At the Power Company village "the millman's store is not a grab-all operated by commercial highwaymen, but one conducted by industrial experts who wish well-fed and well-housed workers." Screened houses, comfortable beds, running water, and shower baths were there already, and more amenities were promised. On top of it all there were "plenty of electric lights." The camp was enough to make a Birmingham miner abandon his picks and head south.[9]

Up in Alabama Power's corporate offices they must have read the *Age-Herald*'s column with particular delight, especially the paragraph that declared how "fortunate are the young men who are in that camp, who have gone from the cities to get in contact with nature afresh." But they also rec-

Supervisors' cottages, Mitchell Dam

(Alabama Power Company Archives)

ognized that the company was providing amenities to keep these "fortunate young men" on the job. Still, the article's conclusion—that "Mitchell dam will be the sooner and better built for this model camp, because conditions there promote self-respect and bring to the fore the sense of decency which is in all, black, white, red, brown, or yellow"—reflected the company's economic and social philosophy: people work well when they are treated well. The *Age-Herald* was not alone in its opinion. A few months later a reporter from the *Birmingham Post,* down to cover the dedication, left with the "impression that the Alabama Power Co., was a pretty human sort of corpora-

Workers, possibly job applicants,
assembled at Camp Mitchell

tion after all," and though he was not as effusive as his counterpart at the other paper, he admitted that the company "certainly looks after its employees in handsome style."[10]

Even with all this good publicity, getting workers into camp and holding them once they arrived was not an easy task. With a turnover rate as high as 46 percent in October 1921 and averaging 31 percent for the year 1922, labor recruiters and foremen had their hands full. Part of the turnover was the natural attrition that took place when one phase of the job ended and another began; for example, the 32 percent in September was due in part to the departure of carpenters who left when the camp was completed, but that same month 229 laborers were reported lost "due in the main to extreme heat, dissatisfaction over the low wage scale, and our practice of dropping employees from the rolls who are absent three or more days." That loss was also to be expected, for that September *was* hot, wages *were low* until the worker established himself on the job, and the company simply *could not afford* to house and feed employees who did not show up when expected. And besides, this was early in the project, before the dance halls and pool rooms were built and before motion pictures and other forms of entertainment were available. These diversions, supervisors believed, would "go far toward making for greater contentment despite the wage scale effect, and greatly reduce our high labor turnover." They were right, for in time off-the-job attractions did help, but still men continued to come and go from the job with unsettling frequency, and a foreman whose turnover rate was out of line with the rest might have some explaining to do. As a result rumors spread of the extremes to which supervisors would go to hire and hold workers. One tale told of how a foreman set a steel trap, caught a good prospect, put him in a cage, and let him out only when it was time for him to work his shift. Eventually, the story goes, the captive became a regular employee, worked for the company for years, and retired with a pension and a limp.[11]

The turnover rate might have been worse if the company had not instituted a system of physical examinations to weed out unfit workers and those who carried communicable diseases. Given when a man applied for employment, it included blood tests for malaria parasites, and even if there was no sign of infection, each employee had to take a required number of quinine capsules just to be sure. In addition a series of typhoid shots was required for supervisory personnel, though apparently not for most other workers, for if all had been forced to take them, the turnover rate surely would have gone much higher. Given the choice "between typhoid inoculations and typhoid fever," one of those whose "arm looked like a pink bologna for several days" after his shot declared "he would prefer chances on the latter." Most workers, seeing the effect on their supervisors, probably agreed.[12]

Efforts by Dixie Construction to keep the workforce healthy also sent physicians and nurses into the countryside to test and treat people "within a radius of several miles of the dam," a practice begun at Lock 12 and expanded at Mitchell. The *Birmingham News* reported how "company doctors

Gambusia pond, Mitchell Reservoir

(Alabama Power Company Archives)

[were] traveling up and down the river persuading residents to submit to malaria treatment at the company's expense" and how "rigorous steps" were being taken to stamp out the mosquitoes that carried the disease. Principle among the preventive measures was the building of ponds throughout the basin and stocking them with *Gambusia*, the top-water minnows of which company officials had learned a few years back and which they hoped would eat the mosquito larvae. Supplied by the federal government, the minnows were put in the ponds to breed, and when the water from the river rose to cover the pond, the *Gambusia* were released into the reservoir to do their work. There was some concern among local fishermen that "the trout and

other big fish in the lake will not be as fond of the minnows as the minnows are of the young mosquitoes." But as far as the Power Company was concerned, it was worth the risk, and it took the minnows "as fast as they [could] be supplied."[13]

Sickness, of course, was not the only thing that threatened the workforce during those first months of operation. Accidents, caused mainly by inexperience and carelessness, took their toll, and fights among men who were not accustomed to working together added a few names to the casualty list. As there had been at Lock 12, there were men in the village who played free and loose with prohibition laws and were arrested for "spreading the joy of Coosa County moonshine throughout the camps." Four of these workers were apparently considered so important to the job that an attorney was retained in Clanton to see to their cases, and the company made arrangements to have their fines deducted from their wages. Dixie Construction made a diligent effort to keep the camps free from illegal whiskey but to little avail. Even after construction was finished and the water was filling the lake, bootlegging remained a big business, and from time to time the local press reported "another liquor raid at Mitchell Dam."[14]

At Mitchell, just as it had been at Lock 12, and just as it was in Birmingham, Montgomery, Mobile, or any other city where men of different races and from different social classes lived and worked together and where they had free time on their hands and money to spend, some were bound to get into trouble. Inhabitants of the "mushroom villages" that had grown up nearby were ready and willing to help workers spend their money and fill their idle hours. Under an agreement with the local sheriff's department, the company policed its own grounds, and if arrests were made the violators were sent to Clanton. In general this arrangement worked well, but a man in jail was a man off the job, so one foreman, afraid to lose his laborers, reportedly ordered them " 'knocked out' to put an end to any disturbance, rather than have them arrested" and taken away. The workers in question were black, but the company's response suggests that the same action would have been taken if white employees were involved. Not wanting to create tension between workers and supervisors, and not wanting this to become a racial issue, the resident engineer personally looked into the matter to see if there was any truth to what was being said. Unable to substantiate the charges, the investigator notified his superiors in Birmingham of his finding. Then he reviewed company policing policies with the local managers and came away from the meeting assured that they would continue to "handle these matters in the way we desire." That way, obviously, did not include "knocking out" workers.[15]

Of course, keeping race relations calm and tension free was to the company's advantage. This result was accomplished, in part, by keeping the races separated and by keeping lines of authority firmly in place when the two groups worked together. What this meant was that when on the job, black workers, who made up almost all of the common labor, took orders and said little. So long as the white supervisors did not abuse their authority, things went smoothly enough. But in one case, a "difficulty" between foreman and worker got out of hand. A "mob," no doubt a white mob, gathered, attacked the worker, and beat the black man "until it thought him dead." Word of this spread quickly through the black camp, and shortly thereafter "a number of negroes left the Dam." With the ranks already shrunk by a flu epidemic that had hit the work force, Dixie could not risk the loss of more men, so the word went out to supervisors that "such occurrences are unfortunate and costly and should not be tolerated." Apparently those orders were followed, for no further incidents of this sort appeared in the reports.[16]

Incidents such as these usually occurred during the first few months of construction, when the men were new, the job was unfamiliar, and the living conditions were close and different. But these cases were few, and considering the size of the project and the number of men involved (778 were on the payroll by the end of September 1921), it is a credit to the operation that there were not more "disturbances." Most of the men, quite simply, were too busy for such when on the job, and after their ten-hour, round-the-clock shifts, most were too tired to get into trouble. Work on the first cofferdam started late that August, and from then until the dam was finished and the gates were closed in early 1923, workers down at Duncan's Riffle had little idle time.[17]

When comparing Mitchell Dam to the one built at Lock 12, company publications noted that the construction methods were "similar in many respects" and that the "plant installation and size of [the] structure [were also] quite comparable." One notable difference was the addition of a new feature called a "backwater suppressor," the invention of Chief Engineer Oscar G. Thurlow. This innovation, which received national attention from members of the engineering profession and considerable coverage in Alabama newspapers, kept down the "standing wave or back roll" that usually developed in the tailrace where the water flowed out below a dam and kept the current from moving smoothly through the turbines and quickly back into the river. Thurlow designed Mitchell so that "water going over the dam is made to sweep the backwater out of the way so that the water that runs the machinery will have a chance to come out freely." Most newspaper readers probably got lost in the more technical aspects of the plan, but they understood that because of Thurlow's invention, the big difference between Mitchell and Lock 12 was that the new dam could produce more electricity, more efficiently, with the same amount of water.[18]

September was the beginning of the dry season and thus the best time to work in the river. Slowly the cribs of the first cofferdam snaked their way from the western bank. Built on shore, then floated into place, sunk, and filled with rocks from the quarry, they looked very much like the "pig-pens" editor Wyatt described in the *Union Banner*. When completed and "unwatered," an area of river bottom "equal to two city blocks in the downtown district of Birmingham" was exposed and ready for construction of the dam to begin. While this was going on and occupying the efforts of about 12 percent of the workforce, riggers were hanging two cableways across the river and getting the hoists ready to move sand, rock, and concrete into place. The other equipment—crushers, air compressors, pumps, drills, etc.—was similar to that used at Lock 12, and the work performed with them was much the same as before. What was different was the power used. Electricity, brought down from the first dam, drove almost everything but the steam shovels and some of the trains and made the project an excellent advertisement for the company's product. Almost every reporter who visited the site noted this advantage, and one from the *Birmingham News* informed readers that "prettier construction work, free from smoke and soot, and with much less noise than might be expected, could not be pictured."[19]

As far as the schedule was concerned, things fell quickly into place. By the end of September the railroad to Cooper was cleared and graded; the log road built to move the timber that could not be floated was completed;

Workmen placing a crib in the third coffer at Mitchell Dam. Note the absence of safety ropes. *(Alabama Power Company Archives)*

and cut and fill work on the improved highway between the dam and Verbena was done. Meanwhile clearing crews were hard at it, and sawmills were working overtime. The first coffer was enclosed in late October, and in a few weeks it was pumped dry and excavation began for the first section of the dam. In mid-November the railroad line was finished, and material began arriving four times a day from the L&N connection. The east and west cable towers were almost completed, and riggers were getting ready to raise the cableways. Finally, the sewer lines were laid, the filter plant was up and running, and the water towers were full. Through it all, "the Lord [was] good to the builders." While "daily rains" fell elsewhere, there were "only occasional showers at Mitchell Dam," and they fell at times when they did not "interfere with the work at all." Everyone, from engineers to laborers, saw it as a good beginning.[20]

The plan they followed called for the erection of cofferdams in three units. The first, and largest, was the one that came out of the west bank. The second would be built out from the east side "to the east edge of the deep water channel." This smaller coffer would be started in late winter, even though the work on the section of the dam in the first coffer was not finished. Therefore, there was a short time when most of the river would flow through the middle section, where the third cofferdam would eventually be built. This was the deepest channel of the river. Flood records kept before Lock 12 gave company engineers a good idea how much water to expect during this critical period, and they planned to build the cofferdams to give the men and equipment inside "reasonably safe protection" from rising waters. But the average rainfall in the Coosa Basin above Mitchell was "greater than any other place in the United States, except some portions of the Pacific coast," so everyone knew the river level could rise higher than their charts told them it would. And if the Coosa rose, there was little they could do to stop it. The project had been lucky that fall, and those in charge could only hope that luck would hold through the winter and into the spring.[21]

On the job with Dixie Construction and the company's engineering department was a resident inspector from the Corps of Engineers. According to the license granted to Alabama Power, the dam had to meet certain federal guidelines, including setting aside an earth space in one side of the dam where a lock might be added if Congress ever decided to open the Coosa for navigation and allocated the money to finance the project. (Most dams built by Alabama Power have that feature, and reporters seemed to mention the Coosa's future as a transportation artery as much as they discussed the hydroelectric potential of the stream.) Alabama was expected to provide the inspector "quarters while in the field," and since he was bringing his wife with him, the company offered a residence like those that housed employees of his status and circumstances—an "unlined three room house with electric lights and filtered water, but without bath." The company told federal officials that its employees furnished "items of food, fuel, ice, house furnishings, clothing, etc., . . . themselves," and it was up to each "to make their homes comfortable to the degree desired by them." All of these things would be the inspector's responsibility as well. Usually, company employees were charged $10 a month rent, which included water and lights, but because it would "require less book keeping," the rent for the inspector was waived. In addition, the company provided the Corps $2,000 to cover the inspector's yearly salary. All the Corps provided, it seems, was the inspector.[22]

The federal inspector was on site when the first cofferdam was closed and unwatered and when the quarry blasting began and the rock crusher went into operation. When the foundation for the first section was drilled, excavated, and poured, he was on hand to approve. He was there in early March of 1922 when, despite predictions and precautions, high water topped the cribs and flooded the first coffer. It might have been worse, if they had not had "the splendid co-operation of those in charge of the operation of Lock 12." More than once Mitchell's managers watched as "favorable operating conditions and the capacity to generate more power were sacrificed [at the dam above them] in order to save [Mitchell's] cofferdam from being flooded." By lowering the Lock 12 pond before high water hit, it was possible to stem the flow coming downstream and reduce the pressure on construction below. In the end the river won and the coffer went under, but still the strategy gave the people downstream time to get equipment out of the cofferdam and "tie down all forms and lumber that might be washed away." One would think forecasters had learned their lesson, but no. In early April the men who analyzed weather patterns declared that "spring floods were over and there was no longer any danger in obstructing the normal flow of the river," so workers started building the second coffer. The weathermen,

River at flood stage, Mitchell Dam

of course, were wrong—twice. On May 5 and again on June 6, despite help from Lock 12, the second coffer was flooded, so it was not until summer finally came and the rains were over and gone that the second coffer was finished and the river flowed through the central opening and through the culverts in the first section of the dam.[23]

By late summer it was time for the third cofferdam. Though the smallest of the three, it was to be built in the deepest, swiftest part of the river, a location made even more complicated by the uneven bed that promised to make a clean seal difficult if not impossible. Aware of these conditions, engineers knew it was all the more important that the work "be far enough along so as to be out of danger from high water before the winter floods set in," so their efforts took on a special urgency. As before, the first cribs were built on land, floated into place, secured with cables, then sunk. Because the water was so swift and the bottom so irregular, double cribs had to be used, and clay was added to the rocks in the fill. Still there was leakage through the "crevices and blind drains . . . in the rocky river bed," and pumps ran full time to give the men a relatively dry place to work. All the while, supervisors watched the sky and worried. River readings and weather reports from Rome, Gadsden, Childersburg, and Lock 12 were telephoned daily to the operating department in Birmingham and then relayed to Mitchell, so if heavy rains fell in the upper basin the men at what was once Duncan's Riffle would have three or four days to get ready. The weather cooperated and the work went well, though, so by the end of August the third cofferdam was finished and ready to be unwatered.[24]

All of this project was captured on film for posterity—and also for propaganda and promotion. When Mitchell was still in the planning stage, the management of Alabama Power "decided upon motion pictures as the best means of recording the construction progress." "Wild charges heaped upon the company" in the past revealed, management believed, "a shocking lack of knowledge" about Alabama Power and its operations. Now they would set the record straight, "tell the people of Alabama the facts, and . . . 'tell it with pictures.' " In addition to educating the public, company officials planned to use the films "to carry the message of Alabama's wonderful industrial opportunities to the people of other states" and in the process "supplement the campaign being waged by trade journals, advertisements, representatives, etc." Complete converts to the use of media to get across a message, corporate leaders accepted the notion that "a more accurate conception of actual conditions can be gained from [motion pictures] than from any other source," because, "the camera does not lie, or even exaggerate." So they hired a Chicago film company and sent it down to Chilton County to make a record of what was taking place.[25]

In fall of 1921 and in May, September, and December of 1922, the cameras rolled at Duncan's Riffle. They filmed the work in various stages of progress, the plant with its turbines, the camp facilities, spectacular blasts at the quarry, men on the job going to lunch, and the Coosa at flood stage. When they finished, the crew went back to the lab to develop the film and edit the

96

results. Meanwhile, other crews were taking pictures throughout the state, and these, when combined with what was recorded at Mitchell, were enough to make three motion pictures: *King Cotton, Electricity at Work,* and *Minerals and Metals.*[26]

The part Mitchell Dam played in these films is unknown. Efforts to find prints of them have been fruitless, and all that remains is an account in the January 1923 *Powergrams,* which noted that "while the use of electricity . . . is shown incidentally, the pictures are in the main devoted to boosting Alabama and calling attention to industrial opportunities." Seen by over 10,000 at the National Exposition of Power and Mechanical Engineering in New York, these films did much to counter "the popular belief . . . that Alabama was a semi-tropical state, with many swamps, cotton fields and negroes, and a few industries." After seeing the pictures, the audience was reportedly impressed instead by "Alabama's advantage over other states in mild climate, abundant raw materials, native American Labor, adequate rail and water transportation facilities, cheap and plentiful water power, low building costs and free building sites and exemption from property taxation in some sections." The film was an invitation for industries to come on down, and *Powergrams'* gentle denials notwithstanding, it was also an invitation to use electricity provided by Alabama Power.[27]

Life at Camp Mitchell

The living conditions at the camp are far above the average on a construction job of this kind. The camp is provided with a modern sewerage system, electric lights, bakery, shower baths, etc., and the workmen are contented.

—*Clanton Union Banner*, February 2, 1922

Mitchell dam will be the sooner and better built for this model camp, because conditions there promote self-respect and bring to the fore the sense of decency which is in all, black, white, red, brown, or yellow.

—RICHARD A. JOHNSTON, *Birmingham Age-Herald*, October 15, 1921

THAT THE ALABAMA POWER COMPANY AND DIXIE Construction treated their employees better at Mitchell than employees were treated at Lock 12 goes without saying. If the company had learned anything from the previous project, it learned that attention to the workers' needs was good business, and as Thomas Martin frequently noted, Alabama Power was a business. But many of the men who built this second dam found more than good working conditions. Because of the way they were treated in the camp and on the job, some who labored at Duncan's Riffle developed a loyalty to an organization that they believed was loyal to them. So when one project was finished they signed on for the next, and then the next, and soon they found themselves permanently employed. Out of this state grew the feeling of family that one senses when talking to retirees who still remember the dams being built and their association with them. "We dam folks stick together," Geraldine Hollis Lawrence told Frank Greene at the party that celebrated his ninetieth birthday. Greene worked those early days at the dams, made his career with the Power Company, and finally settled in Chilton County on the road leading to the dam. Lawrence, who lived in Clanton, grew up in the Mitchell Dam village, where her father was superintendent. She and Greene had been friends for years. Dam people did stick together.[1]

Hospital ward, Camp Mitchell

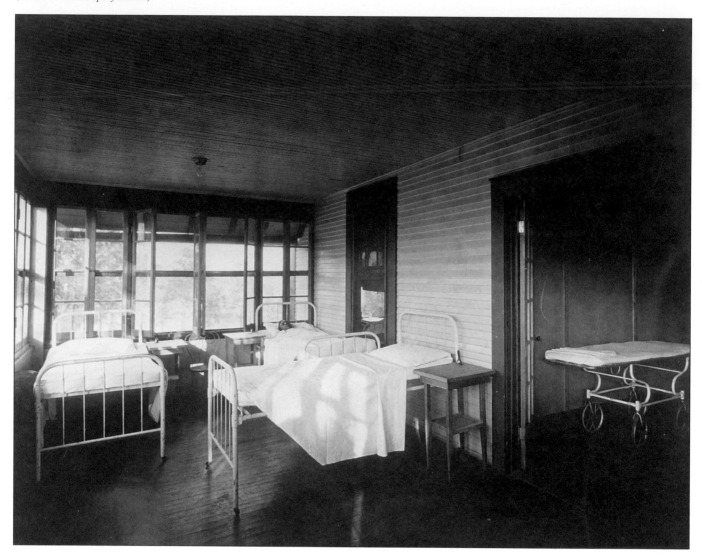

■■■

Of all the changes the company made to improve the lives of its employees at the site, the most sweeping (and in many ways the most appreciated)
concerned health care for the workers and their families. One of the new
aspects of the Mitchell camp that set it apart from the one at Lock 12 was a
modern hospital on site. A physician who was also an Alabama Power company employee was in charge of the facility, and he was given sufficient staff
to serve the residents of the reservation. These two innovations reveal once

again that the company had learned valuable lessons earlier and knew what
to avoid as well as what to repeat. No longer would there be a doctor working
for himself at the site, cutting corners at the expense of the employees; and
no longer would cases go untreated, or poorly treated, because the physician
and his staff lacked the facilities to do the job properly.[2]

Dr. Samuel R. Benedict was appointed chief surgeon, and under his direction health care at Mitchell Dam became a source of pride for the company. A southerner with extensive experience in industrial medicine, Benedict had been elected vice president of the Association of Seaboard Air Line
[Railway] Surgeons and had served on a special commission that investigated "malaria and slow fever" in south Georgia. Well aware of what needed
to be done, he made the construction of a hospital his top priority, and when
some in management suggested that with labor short some carpenters employed there might be transferred over to the cofferdam, the doctor fired
back that *because* of the labor shortage, work on the medical facility should
move ahead as rapidly as possible. With a hospital available, he argued, "it
will not be necessary for any men to lie up in the bunk house sick" where
they would require the "attention of others, which is a direct loss, and to a
certain extent demoralizing to the other employees." And if that opinion
was not enough to move those in charge, the doctor added that "under the
Compensation Law, the only way in which we can protect ourselves is to give
the employees who are injured the best attention possible and return them
to work in the shortest period of time with the least amount of disability."
Dr. Benedict knew which buttons to push to get things done, for company
officials remembered well the lawsuits brought over injuries at Lock 12. The
carpenters remained on Benedict's job, and the hospital opened on December 17, 1921.[3]

Not long afterwards *Powergrams* announced that Alabama Power and
Dixie Construction, "realizing the responsibility of taking care of the employees and their families," had built at Mitchell Dam what was, in all prob-
ability, the finest medical facility in the region. Large and attractive, designed in the English cottage style popular at the time, the hospital was
laid out to accommodate fifteen patients, though more could and would be
served at times. There were, of course, separate wings for the races, and in
these were separate wards for men and women, for company planners now

Visitors from the Medical Division of the
League of Nations at Mitchell Dam, 1923

(Alabama Power Company Archives)

medical community. In the fall of 1923 "a party of the world's most famous
physicians who [were] touring the globe with the purpose of making the best
methods of disease prevention international" arrived at Mitchell to inspect
the site and hear Dr. Benedict explain the techniques he used to keep the
spot healthy. Part of the Medical Division of the League of Nations and spon-
sored by the Rockefeller Foundation, the group, so the ever-effusive *Powergrams* claimed, "counted their visit to the United States incomplete without
observing" what Benedict and the Power Company had done. Everything
about the visit was a success. The doctors were "plainly delighted" by the hospital and its equipment, they "manifested high interest in the methods used

to foil the anopheles or malarial mosquito," and they took particular note of how the company had stocked the lake with *Gambusia*. Other efforts—the vaccination system for typhoid, the emphasis on pure water for the workers, the arrangement of dwellings, and the sanitation measures—drew their approval. They also enjoyed the guest house and especially the meals served there. The visit to Mitchell Dam was, according to *Powergrams,* considered "one of the most important features of their trip."[26]

This recognition was not the only one given Power Company employees for their efforts. The next year Dr. Benedict was elected a member of the board of directors of the Association of Industrial Physicians and Surgeons at its annual meeting in Chicago, where he addressed the body on the problems facing health officials in the South. Commenting on his speech, the director of the health department of the *Chicago Tribune* called Benedict's contributions "the greatest piece of malarial control work done in the South at any time" and went on to declare that Alabama and Alabama Power Company "are to be congratulated upon the completion of this project, leaving as it does better sanitary conditions in the reservoir than existed previously." Benedict returned shortly to organize and lead the "Association of Power Company Surgeons," made up of physicians from the towns in which the company operated. The plan was that these "eminent medicos" would meet from time to time to exchange ideas "in a way that is decidedly beneficial to them and to the Company whose medical safety and sanitary problems they are solving." One of the first places the group gathered was Mitchell Dam.[27]

All these honors notwithstanding, Mitchell Dam was really about electricity. With its capacity to produce power greatly increased, the company moved into new markets, and in the spring of 1923 the city of Montgomery signed up to get its electricity from the Alabama Power Company. But not every request for "juice" was as welcomed. That same year A. L. Robinson, who lived not far from Mitchell Dam, wrote the company that he wanted power as well. Robinson, who had been "some trouble" when the dam was being built, claimed that when he sold Alabama Power some of his property he was told by the agent that he would have lights when the job was finished. Convinced that no such promise was made and that Robinson's memory was faulty, J. M. Barry, who managed the company's retail operations, contacted the superintendent at Lock 12 to see what could be done. The answer was not much, unless Robinson wanted to pay for "a costly extension" from the Mitchell Dam hospital to his home. Alabama Power might help him with this expense, it was suggested, if he was willing to "increase [the company's] revenue by agreeing to install an electric range." Apparently the deal was put to Robinson, and he rejected it. Like so many rural folks, he would have to wait his turn for the lights to come on. But in time they would, and his life would be brighter for them.[28]

Finally, a Dam
at Cherokee Bluffs

The "city" of Cherokee Bluffs will be a larger one than Tuskegee,
Sylacauga, Marion, Alexander City or Woodlawn. Past experience has
taught Company engineers exactly what is required in a community of this
sort and though the one at Mitchell Dam was considered almost ideal, they
have stated that experience there will enable them to make an improvement
in the one at Cherokee.

—*Powergrams,* July 1923

CHEROKEE BLUFFS WAS ONE OF TALLAPOOSA
County's most famous landmarks. The scene, according to the editor of Camp Hill's *Tallapoosa News,* was "one of grandeur," and
people from all around the region knew that it was "worth the effort of anyone to go and see it." Guesna Neighbors Moon, eighty-seven years
old, recalls how Sunday school classes would go out and stay there overnight.
Her father would go along and cook for them. Supper was easy to come by,
for the river teemed with fish. John D. Towns, another longtime resident of
the area, remembers how back before the dam fishermen pulled giant catfish from the deep holes between the rapids. One day a local fisherman came
up from the river carrying a fish so big that its "head was just at [the man's]
shoulder and the tail of the fish was touching the ground." But it was not
grandeur or fishing that attracted James Mitchell. Upriver from the bluffs
the Tallapoosa's banks grew tighter and its current swifter, and at Cherokee
this constriction reached its narrowest point. There the stream rushed over
a bed of "solid rock" and between walls of granite that "extended abruptly
downward to the edge of the water a perpendicular distance of 160 feet."
The surrounding country was "hilly," leaving farmers only narrow river bottoms and small ridges to cultivate, so the population of the region was scattered and sparse, as it had been over on the Coosa. Such land could be
bought cheaply and quickly, so the "overflow cost [would be] practically

117

Fishermen at Cherokee Bluffs, shortly before
work on the dam began

(Alabama Power Company Archives)

nothing." Given all these advantages it is hardly surprising that Cherokee Bluffs was James Mitchell's first choice for a hydroelectric facility. It was a perfect place to build a dam.[1]

James Mitchell was not the first to see the potential of the site at Cherokee Bluffs. Not long before he arrived two other pioneers, Henry Horne of Macon and James R. Hall of Dadeville, had explored the area, looking for dam sites and securing options. Their guide and ally in this was Mrs. Nora E. Miller, a Tallapoosa native of "wide acquaintance" who knew the country and was held in "great esteem" by local residents. Mrs. Miller, who was described as a woman "fitted both by heredity and upbringing for the active part she took in the affairs of her section and state," listened to the two men and became a convert to the idea of developing the Tallapoosa. Horne and Hall failed to get the financial backing they needed and abandoned the project, but Nora Miller's enthusiasm never waned, so when James Mitchell appeared on the scene, she embraced his cause. She invited Mitchell to stay in her home, and together they "explored the river in [her] big car." When they completed their investigations, James Mitchell had a site for a dam, and "the plans for the river's development . . . had been formulated in his mind."[2]

That happened in 1911. The next year Mitchell and his new partner, Thomas Martin, took control of the Alabama Power Company, and the development of Cherokee Bluffs was at the top of their agenda. Survey parties were sent out to do preliminary work at the dam site and furnish engineers with complete topographic data for the area. Land agents went out as well, and once again Mrs. Nora Miller was the company's contact. With her help Alabama Power secured "several very vital locations on the river without which the great developments contemplated could not have proceeded," and for a while at least it seemed that Cherokee Bluffs would be the Alabama Power Company's first hydroelectric project. But, as already noted, the project was slowed and then halted by interests such as Mount Vernon–Woodberry Cotton Duck Company, so Mitchell and Martin turned their attention to the Coosa and the Lock 12 site. That dam was finished, and the one at Duncan's Riffle was well underway when the last hurdle was cleared, and the development of the Tallapoosa could begin.[3]

■ ■ ■

On September 1, 1922, the *Alexander City Outlook* announced to its readers that Alabama Power planned to build not one but "4 Dams on [the] Tallapoosa" to meet the growing power needs of the state and region. Of more immediate importance, however, was word that the dam builders would hire and buy from the immediate area. Alexander City, Dadeville, Tallassee, and all the other towns, villages and crossroads in the area knew what had happened to towns near the Coosa. Alabama Power had come to Clanton, Verbena, and Cooper and brought with it money and jobs. Residents around the Tallapoosa decided it was about time that they got their share.[4]

The *Outlook*'s announcement was a bit premature. Federal and state re-

hospital [was] closed" ahead of schedule. With men like Davis on the payroll, Dixie Construction could boast of a first-rate management team.[19]

■■■

With so much clearing to be done, everyone who wanted a job could find one, and workers descended on Tallapoosa County well before the camp was built to house them. Some found room and board in local homes, where they were charged a dollar a day, including breakfast and supper. At times the arrangement worked so well that some boarders stayed even after the camp was built, and at least one of them married the daughter of the landlord. Others lived in "little houses" quickly thrown up and rented by the same sort of entrepreneurs who built the boomtowns near Mitchell and Lock 12. Still there was a housing shortage. Soon farmers were no longer surprised to come into the barn at daylight and find workers who had spent the night in the loft. In the towns men on the way to the site slept wherever they could find a vacant building, an open door or window, and a dry spot, which is why the city fathers of Dadeville passed an ordinance that prohibited people from sleeping in church. A few local ministers probably wondered if they could enforce the rule on Sunday mornings.[20]

Meanwhile planners laid out the camp, and carpenters went to work. Based on what was learned at Mitchell, management believed it knew "exactly what is required" to build "a modern city in the wilderness," one with "all of the necessities and many of the luxuries usually found" in similar communities. When it was finished the "city of Cherokee Bluffs," with a population that may have gone over 3,000, would be the Tallapoosa Valley's largest town—bigger than Alexander City, Dadeville, Sylacauga, or even Tallassee.[21]

On the whole the camp on the Tallapoosa was very much like its predecessor on the Coosa, only larger. Located on high ground on the east side of the river, it covered 117 acres and eventually contained 332 structures of various kinds. Although the initial purpose of the camp was to provide housing for construction workers, buildings "were laid out so that they could be converted later to the uses of the permanent operating forces." As at Mitchell, the best accommodations were set aside for superintendents and their families, but there were well-appointed foreman quarters in the white camp, as well as nine bunk houses for white laborers. There were also an

Dance hall for black workers,
Martin Dam

(Alabama Power Company Archives)

amusement hall, a two-room school, a church, and a pool room—something that had been available only in the black camp before. There were two mess halls, one in each camp, and the bakery and ice plant were close enough to serve both. A barber shop, shoe shop, and garage were also among the facilities, as was the commissary. The streets and buildings were, of course, lighted by electricity.[22]

The "negro camp" also reflected the company's experience at Duncan's Riffle. Instead of crowding most black workers into large bunkhouses, Dixie constructed some 200 "one-room shacks," which *Powergrams* proudly described as "houses of the sort the dusky laborer prefers to any kind of habi-

through, those parts of the basin that would be exposed when the water fell were "clean as though it were to be used for cultivation." Meanwhile, using Mitchell as the guide, potential employees and their families were tested for the parasite before they were allowed in the camp, and they were rejected if malaria was found. Those hired were given regular doses of quinine during the mosquito breeding season. And as in the past, these services were available free to people throughout the basin.[13]

The company remained convinced that "larvae eating minnows" like those put in the lake at Mitchell would help keep down the mosquito population at Cherokee Bluffs as well, so twenty-five "*Gambusia* pools" were built throughout the basin. Ranging from one-eighth to two acres in size, they were stocked with nearly 50,000 of the small fish that would be released when the water rose and the reservoir filled. These pools were an important part of what *Powergrams* described as the "most extensive program for malaria control that has ever been put on in the Southern States." John Towns's father was the man responsible for inspecting the pools to see "if the [water was] healthy and the fish were growing properly." Riding the "beautiful saddle horse" provided by the company, Inspector Towns went from site to site looking for beavers that might dig holes in the dams or checking to see if recent rains had flooded the ponds. On some trips, the ones that were not overnight, young John would ride horseback on the rounds with his father. The basin was still being cleared, and he remembers the men at work and the fires from the brush being burned. Even a young boy could see and understand that the dam at Cherokee Bluffs was the biggest thing ever to come to Tallapoosa County.[14]

Had John Towns and his father gone downstream to the site itself the boy would have been even more impressed. With the railroad completed by early 1924, men and material were arriving on a regular basis, and the operation took on the look of an army camp on the eve of an invasion. In a way that was just what it was. The enemy was the Tallapoosa, and Alabama Power was rallying its forces to defeat it. Lloyd Frank Emfinger was soon promoted from his "powder monkey" job and put in charge of a crew digging test holes and wells. He and the others used a well borer to cut through the rock to a depth of around sixty feet. Then they packed the hole with black powder and just enough dynamite to set off the charge. Wires ran from the plunger

Early blasting work at Cherokee Bluffs site

(Alabama Power Company Archives)

to the capped explosives, and when all was clear Emfinger pushed the lever, and "it went off."[15]

■ ■ ■

J. F. Fargason was among those who came to the Bluffs to watch Emfinger at work. When "word got out that they were going to blast thirty-three cases of dynamite at one time," he and his friends "got a crowd and went down to see it." The company needed a lot of rock for construction, and since the engineers wanted to get an outcropping below the dam "out of the way," the blasting crew was sent down and told to move it. Fargason and his friends took their place on the hill near the commissary and waited for the explosion. "We didn't know what to expect [so] everybody had their ears stopped up." Then it went off. At first they were disappointed. "It wasn't a very loud

Thomas Martin, on platform, speaking
at cornerstone laying, Cherokee Bluffs,
November 7, 1925 (*Alabama Power Company Archives*)

as well. "The continued progress of our State," he told the crowd that day,
"consists in lifting the burdens of drudgery from the shoulders of man to
the tireless shoulders of the dynamo. Every loafing stream is loafing at the
public expense and every kilowatt of power means less work for someone,
more freedom and a richer chance for life." Thomas Martin and Alabama
Power were committed to putting loafing streams to work.[22]

So it followed naturally that about seven months later, one week after the
gates were closed at Cherokee Bluffs, the board of directors of the Alabama
Power Company resolved that the "development hereafter be named and

designated as 'MARTIN DAM' and that the great body of water to be formed thereby be named and designated as 'LAKE MARTIN.' " The *Birmingham News* claimed credit for the idea, pointing out that "it seemed only right and fitting that this wonderful lake and piece of engineering should bear the name of the constructive genius who captain[ed the Company]." If anyone disagreed, they had the good sense to keep quiet, so the board "enthusiastically" (according to *Powergrams*) made the change. The resolutions also called for a "fitting ceremony" to be held and a "suitable plate or tablet be placed giving proper recognition to the work of Mr. Martin in the interest of his state," but Martin thought that such a ceremony honoring the president of the company might seem self-serving. So they took a lot of pictures and waited for his consent to place the plaque. Ten years later he finally gave in.[23]

During all this activity at company headquarters, work on the dam continued. Now entries in the superintendent's journal noted more things being dismantled than built. Out in the basin, acres of second-growth brush had to be cut and burned, and clearing crews worked frantically to beat the slowly rising waters. At this point the Mexicans were brought in. It was harvest season, picking time, so local workers were tied up on the farm and there was a critical labor shortage. Where the Mexicans were recruited, how they got to the site, and what they did after they arrived remain a mystery, and were it not for the memories of a few retirees and a couple of references in the records, no one today would know they were ever there. The Mexicans were "housed in shacks and bunk houses originally built for negroes," and they apparently did the "dirty and nasty" work that was a necessary part of the final phase of the operation. For the most part they kept (or were kept) to themselves. They did not get along well with blacks who were still on the job and created some problems for local law enforcement on payday. When the work was finished they apparently returned to Mexico, leaving little in Alabama to indicate that they were ever there.[24]

Meanwhile Alabama Power and Dixie Construction had one more loose end to tie up, and it was a big one—the Kowaliga Creek Bridge. From the beginning local officials were concerned at the impact the lake would have on local transportation. Over a hundred miles of roads would go under the water, and travelers would be faced with a wide lake where once there had been easily forded or ferried streams. Kowaliga Creek was one of these. Maps revealed that the reservoir would push water up the creek and flood the bridge on the busy Alexander City–Tallassee road, leaving travelers to find a way to cross a stretch of lake more than one-half mile wide and ninety feet deep. Alabama Power realized the problem and "recognized its responsibility in the matter," but when it proposed to establish a free ferry, local residents would have none of it. Commissioners in both Elmore and Tallapoosa counties got wind of the matter, and soon Kowaliga Creek became a political issue. Public hearings were held in Alexander City and Dadeville, and the whole thing seemed on the verge of becoming a public relations nightmare. So the company did what it had to do; it agreed to build a "modern highway bridge of a permanent nature" across the wide expanse of water that was once Kowaliga Creek.[25]

Agreeing was one thing—building quite another. Engineers examined the site as it was being cleared and concluded that the best thing to do was "to erect the foundations, piers and trestle . . . before the filling of the reservoir was started and to install the steel girders and decking after the reservoir is full." According to this plan, the girders, each 111 feet long, would be loaded on a barge at the dam, then floated across the lake and maneuvered into position. Getting everything into place was an operation that required the skill of a master "rigger" and more than a little seamanship, but Power Company engineers were sure it would work. Under the watchful eye of C. C. Davis, work on the piers and trestles began in November 1925. Only seven months were left before the dam was supposed to close, and the rainy season was approaching, so everyone knew the schedule was tight.[26]

Of course they made it. By the time the water began pushing up Kowaliga Creek, the piers, some of them over a hundred feet high, were in place, and the girders were being assembled down at the dam. Finally, in May 1927 the lake was high enough to begin moving the girders to the barge and floating them to the site. It was a tricky operation. The I-beams that would carry the girders to the barge were greased with beeswax so the load would slide easily aboard, and once they were in place a towboat would pull the cargo slowly across eight-and-one-half miles of lake. There were twenty-four girders to be carried, and four could be put in place in a week if the weather cooperated, which naturally it did not. On one occasion a storm blew boat, barge, and

Kowaliga Creek Bridge under construction

girders some two miles off course, where they stuck fast on a stump. The load had to be shifted to get it off, and one miscalculation could have sent the whole cargo to the bottom of the lake; but they got it free, and soon the operation was back on schedule. By the end of June the girders were in place, the concrete floor was being poured, and arrangements were being made for the Kowaliga Creek bridge to open in August. It had cost nearly three-quarters of a million dollars to complete. Little wonder the company would have preferred a ferry.[27]

By then work at the dam was also finished, and Dixie Construction had turned operation of the powerhouse and other facilities over to Alabama Power engineers. But before that change in management was accomplished, the mighty machines at the facility had to be put into operation and the dam

by Elmore County, these men enforced the law in and near the camp and from most reports were able to maintain order with little trouble. In fact, one former employee recalls that residents of the village were pretty peaceful and that altercations, when they did occur, usually took place outside the reservation between men who either lived nearby or walked a distance to settle their differences. Few, it seemed, were willing to create a disturbance at the site and risk losing their jobs. On the whole workers were usually too tired after a ten-hour shift to cause much trouble, and, as the company discovered at Martin, families in the camp helped keep order as well. What Alabama Power created at Jordan Dam was a village *and* a community.[12]

The main focus, however, was the job. By June 1926, the railroad was running, and supplies were piling up on the west bank. Applicants arrived and were given their physicals; if they passed they were assigned to quarters and put to work. Some of these were mighty men, and stories of their exploits are still handed down. Jack Mitchell was one of them. "Strong beyond belief," according to Howard Bryan, Mitchell one day was carrying a cross tie from one part of the site to another when C. C. Davis stopped him. "You going to hurt yourself," Davis told him. "Next time you go down there double up." So Mitchell went down, and when he "came back the next time he had one under each arm." Others were noted for their colorful language, like the man who could not put together a sentence without cussing. If it "had as many as three words in it," Bryan recalls, "one of them was going to be profane, probably two." And there were men like "Peavine" Miller, who checked lines that ran out from the site and of whom it was said "if he walked up on a tent meeting he'd get religion, if he walked up on a still he'd get drunk." Yes, those were mighty men.[13]

And there were a lot of them, or so it seemed. To B. K. McDonald, the work "wasn't that hard because you had enough help." Whatever needed to be done, Dixie Construction sent more than enough men out to do it. "Nobody was put in a strain." McDonald started at Jordan Dam just as it was coming on line. He stayed with the company for thirty-six years and in that time "never knew of anybody asking you to do more than what you should do." Looking back he could say, "I loved every day." McDonald's sentiments were not shared by everyone. The workers who had to be "rousted" out in the morning to get them to the job probably felt they were being asked "to do more than what [they] should." Some of them even took to hiding in the outhouses, but the rousters found them anyway and docked them an hour for malingering. Employees especially disliked the night shift, and it seemed to McDonald that Dixie "had to hire twice as many laborers as [it] needed in order to have a crew on the job" after dark. But if the dam was to be completed on schedule, men had to work around the clock, so they did.[14]

They also played. Perhaps it is because Jordan is a later dam, and therefore there are more folks still around to remember life in the camps and on the job, that there are more personal accounts of what things were like back then. And because people cling best to warm memories and remember even the bad times as good, there are more stories about amusements than about work. Sometimes the job itself was entertainment—like the time the "gandy dancers" arrived. They were part of the crew that worked the railroad, black men who "put down the track and picked up the track," who laid it straight and who trued the curves. B. K. McDonald remembered they had a leader who was hired just to "sing cadence . . . and he would sing yaddi, yaddi, yoddi, yoddi, yo ho, and that's when they'd lift, all go at the same time." Then he'd start singing again, "and he'd get to a place where there'd come a lift and out they'd come with it," right on cue. They called the song "chanting for the lifts," and when they worked, folks stopped to watch.[15]

The camp at Jordan provided many of the same opportunities for company-sanctioned amusements as the camps at Mitchell and Martin, though with Wetumpka so close by, it seems that a greater effort was made to keep black workers in the camp and not abroad in the countryside. In addition to the poolroom, which was company operated, Dixie built a thirty-by-fifty-foot Negro dance hall. It was operated by a black businessman, who was also allowed to sell tobacco and soft drinks when the pool room concession stand was closed. Soon after the dance hall opened, the manager "installed a piano and employed a player," and before long this establishment became one of the most popular in the camp. The white section of the village was even better supplied. It had a thirty-six-by-seventy-two-foot "community hall," where motion pictures were shown once a week (westerns were a great favorite). On the Sabbath the faithful used it for Sunday school and church. Folks coming in on Sunday morning probably had some cleaning up to do, for among the uses to which the community hall was put, Saturday night dances were

168

get dressed, take a jug of whiskey out from under the bed where he always kept it hidden, and wake "Mac" to join him in a "stiff drink" before he'd leave. McDonald, who did not drink much, usually declined, so his room-mate would slug it back and head out into the night. With whiskey so easily available, some might have wondered why law enforcement did not crack down harder. All sorts of rumors gave reasons, but among the most popular was the belief that the "sheriff can't catch the bootlegger because he *is* the bootlegger." That may not have been true, but it was a popular business, and many folks were involved in it.[18]

Stories like these, filtered by time of any impurities that might taint the product, leave the distinct impression that life "was good" in the camp. This was especially true in the bachelor quarters, where former residents count among their advantages the "good food" prepared by a black man and his wife. Known for preparing "the best breakfasts that you have ever seen," the cooks also provided the fixings for sandwiches that the men could take out for lunch on the job ("we ate a lot of bologna" one recalled), and when the day shift came in, the men were treated to an evening meal that was as good as breakfast. If after supper they were too tired (or full) to go into town, they could hang around camp, shoot a little pool, see a movie, or take part in "some of the biggest poker games" one could imagine. Yes, life was indeed good.[19]

Of course not everything was ideal, as the hospital records clearly show. This job, like the jobs before it, was dangerous. In the twenty-nine months that medical records were kept, the doctors and their assistants treated 2,560 accidents. Over 10 percent of the men on the payroll were injured in some manner. Add to that the over 9,000 times the doctors treated sick workers and members of their families for various ailments, and Camp Jordan appears to have been neither a safe nor a healthy place. Closer examination suggests that quite the opposite was true. On the job was a full-time "safety engineer" who "was occupied with accident and infection prevention work and personnel matters." He was aided by the "first aid man" who "treated injuries of a minor nature" at the "first aid-house," a new addition that was part of the receiving station where prospective employees were examined. His job was to look after simple cuts and bruises, then send the men back to work. Most of the accidents—abrasions, lacerations, contusions, foreign bod-

Hospital, Jordan Dam
(Alabama Power Company Archives)

ies in the eyes, and "miscellaneous pains"—fell into "minor injury" classification.[20]

Still, one should not minimize the risk involved in work on a hydroelectric dam. Conditions that would make modern safety inspectors cringe were considered a natural part of the job. Eleven men were killed working on Jordan Dam, five during the first seven months when most common laborers were new to the job. Doctors at the hospital performed thirty-four surgical procedures ranging from hernias to amputations, all of them successful. In fact, the hospital is remembered as one of the best-run operations at the site. It was "located on a gentle slope at the edge of the permanent camp and accessible to all parts." Hospital grounds were "beautified by flower beds, shrubs, and grass"; there was an "excellent vegetable garden" nearby to provide patients with fresh produce. With wings for the two races, the building

Girls in knickers, Lake Martin

(Alabama Power Company Archives)

and frequently living in boarding houses that catered to single women, and they responded enthusiastically to the opportunities it afforded. In the summer of 1924, a party of thirty-two "merry girls, most of them attired in their knickers," left Birmingham on a Friday afternoon. The Benzingers met them at Ocampo and took them to the lake, and with "the silvery moon furnish[ing] light by which to sing and dance," they floated down to the dam. "While it is difficult to improve on nature," one of them wrote after seeing the Lock 12 facility glowing in the dark and reflecting on the water, "it can be done at night very effectively with the aid of electric illumination." That night was spent at the guest house. The next day they were entertained at Camp Mitchell; then they returned to Lock 12 and home on Sunday. The trip was a total success, and those who did not go were told to "save your pennies" for next year. Apparently many absent employees did as they were advised, for in 1925 the knicker-clad group was "65-strong."[9]

Folks in local communities welcomed the campers when they dropped by on their way to the lake, though the knickers worn by three of the "girls" created "quite a sensation" in Calera, and the town was "still ringing with

A "9-foot, 400-pound sturgeon"
taken from the river below Mitchell Dam

(Alabama Power Company Archives)

talk" when they left. Upchurch's drug store in Clanton, where you could "get a warm reception but cold drinks," became a regular stop on the route, and other merchants also made a special effort to cater to the visitors. Out at the dam villages the permanent residents also looked forward to the "house parties" and excursions, for they were social events for them as well. The attention these guests paid to the village residents, the interest they showed in the work being done, and the praise those big-city folk heaped on the natural beauty of the surroundings gave dam employees a sense of being one with the larger organization, so they welcomed them with open arms—including the "flappers" who "scandalized" their country neighbors.[10]

Of course not everyone went to the lakes to enjoy the "outdoor life deluxe." Lock 12 and Mitchell attracted fishermen from all over the state, and the *Birmingham News* announced that the company planned to build a fish hatchery, the first in Alabama, to help stock the two bodies of water. A report that a man fishing below the Lock 12 powerhouse caught a blue cat so large that a "grocer bought it . . . and sold it out in steaks" encouraged other anglers to try their luck. The record catch, however, must have been the "9-foot, 400-pound sturgeon" that was pulled from the water below Mitchell and displayed for the *Powergrams* photographer. "Why go to the coast," the caption read, "when there are man-eaters such as this . . . awaiting" in the Coosa. Over on the Tallapoosa, Lake Martin became a sportsman's mecca as well. A "special species of carp, previously unknown in the area," appeared in the new lake, and soon locals were coming home with strings of fifty and a hundred. Not surprisingly, this overfishing took its toll, and the carp "gradually disappeared within a few years . . . and now one is rarely seen." But catfish, bass, and bream thrived there, and fishermen came to the lake to catch them. A guest house in the village at Martin entertained its share of visitors, but it would be years before that lake would even begin to become the attraction it is today. Jordan, being close to Montgomery, drew many weekend fishermen and campers, who came out with tents to "rough it" for a few days or who had company connections and were invited to stay in the guest house there. But the showplace was Camp Mitchell, and the people who enjoyed it most were employees of the Alabama Power company—the members of the club.[11]

It followed that local residents gained many advantages from having Alabama Power come into their communities and stay. New roads and bridges were built to replace those flooded out, and the infrastructure of the surrounding counties improved significantly. New businesses started up, like the one begun by the Dadeville entrepreneur who put a "big 40-passenger boat" on Lake Martin, had it christened by the United Daughters of the Confederacy in a "special ceremony," and began running excursions of eight, ten, and forty miles. Fishing camps sprang up along the lakes, and some farmers whose fields were flooded built cabins on the sloughs and rented them to fishermen who came in from the cities. Company efforts to keep the mosquito population under control made the visitors more comfortable in their primitive accommodations and also gave summer jobs to students who sprayed the inlets with oil and did mosquito surveys at residences near the water. These efforts, plus the work of the *Gambusia,* did not eradicate the mosquito or end the threat of malaria, but they made the various locations healthier than they had been before the water was backed up, and that was no small accomplishment.[12]

Of course, the greatest advantage people enjoyed was electricity, the reason the dams were built in the first place, and the company took every opportunity to impress on the public the worth of its product. The public had little trouble seeing the benefits. Prior to 1912 only seventy-two Alabama communities had electricity, but by 1928, when Jordan Dam went into operation, Alabama Power served four hundred twenty-one communities in sixty-one of Alabama's sixty-seven counties. The company also provided power for coal and iron ore mines, cotton mills, cement plants, quarries, steel plants and rolling mills, foundries, pipe plants and machine shops, ice plants, public utilities, and electric furnace installations, industries that put thousands of citizens to work.[13]

Alabama Power lit up the lives of countless Alabamians. Meager records make it impossible to know how many individuals actually had electricity prior to 1912, but apparently the number was small. However, once Lay Dam went on line and Alabama Power began expanding its generation and distribution facilities, the number of consumers grew rapidly. By 1919 the company had over 10,000 customers, and that number had nearly doubled three years later. Then Mitchell Dam began operation, and before 1923 was out over 35,000 were being served. That same year marketing efforts broadened when the company, "with associated local firms," opened Alabama's "first electrical home" in Selma, and citizens saw what the future held in store. They liked what they saw, and by 1926 over 60,000 had signed up. The number continued to rise until the end of the decade.[14]

Most of this expanding market, however, was in the towns and cities, which presented a public relations problem for the company since most Alabamians still lived on farms. The charge that Alabama Power was ignoring the needs of the majority was a sensitive issue at corporate headquarters, and by 1923 the company had worked out an arrangement with Alabama Polytechnic Institute at Auburn to support research "to find practical means and methods of supplying electric service to the farmers." It was recognized that in order for Alabama Power to make a profit on rural service, farmers would have to be shown how they could use electricity for purposes other than lighting. The electrification of the agricultural experiment station was

Distribution system of Alabama Power Company, existing and proposed, after Lock 12 came on line, 1914

(Alabama Power Company Archives)

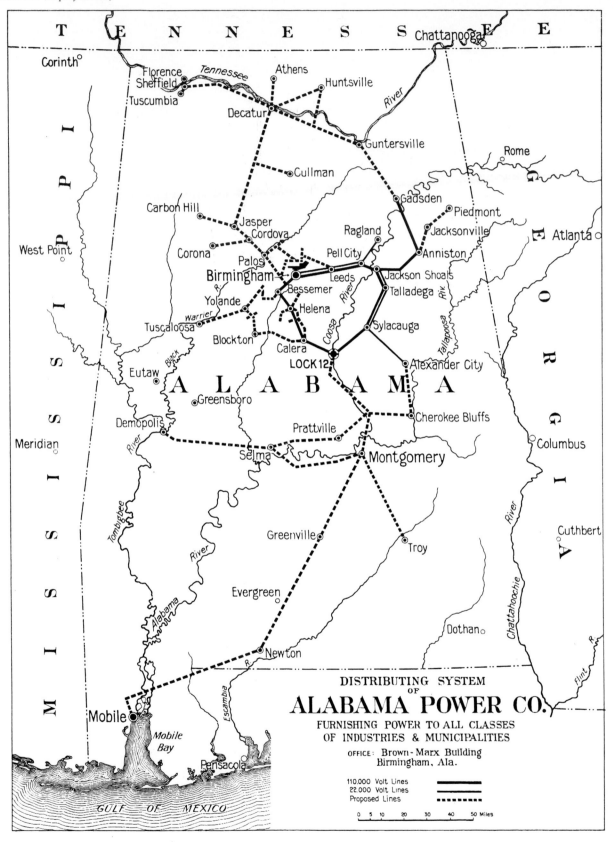

the first step, and in each subsequent *Annual Report* the company told of the progress being made. Slowly but steadily more rural lines were added, and by 1926 Alabama Power was serving over 6,000 agricultural customers. It was, however, a small start, and company officials knew that electricity would not "be a cure-all for the various ills which at present afflict farm life." Still, Alabama Power believed that "the intelligent distribution and use of electrical power on the farm should lessen drudgery, make farm life more attractive and increase production per man, thereby increasing wealth and aiding the reduction of tenant farming." Country folks were willing to give it a try, if they could get a line to their houses and prices they could afford. But for many that would be a long time coming, and Alabama farms would continue to be lit by the glow of kerosene lamps and lanterns.[15]

By the end of the 1920s Alabama Power had transformed the face of Alabama, but in the less than two decades since it came under the control of James Mitchell and Thomas Martin, the company had been transformed as well. Alabama Power was now a mature corporation, with thousands of employees and with plans for the future, but it would be a long time before dreams of more dams on the Coosa and Tallapoosa would become a reality. In 1929, just as Jordan was coming onto line, the American economy collapsed. Projects already financed were carried out, but most had to be shelved. Years would pass before Alabama Power hired hundreds of construction workers again. The depression took its toll at all levels. Men and women in the offices were laid off as the company cut expenses to the bone, while down at the dams operations were carried on with skeleton crews. The APC Club was also a victim of the depression, as those who had jobs held onto their money and stayed close to home. Now the villages did become little worlds unto themselves, where residents rallied to help each other weather the crisis. They still held dances from time to time; the schools still operated; there were occasional parties and picnics, especially for the children. But visitors were fewer now, and the entertainment for them was less lavish. Still, those who lived there had jobs, nice houses, and amenities that their counterparts in Calera, Verbena, and Clanton lacked. Little wonder that, looking back, people who lived out in the counties thought Alabama Power employees were among the most fortunate people in the state.[16]

If citizens from outside the reservation were a little jealous of "dam

Distribution system of Alabama Power Company, existing and proposed,
after Jordan Dam came on line, 1929

(Alabama Power Company Archives)

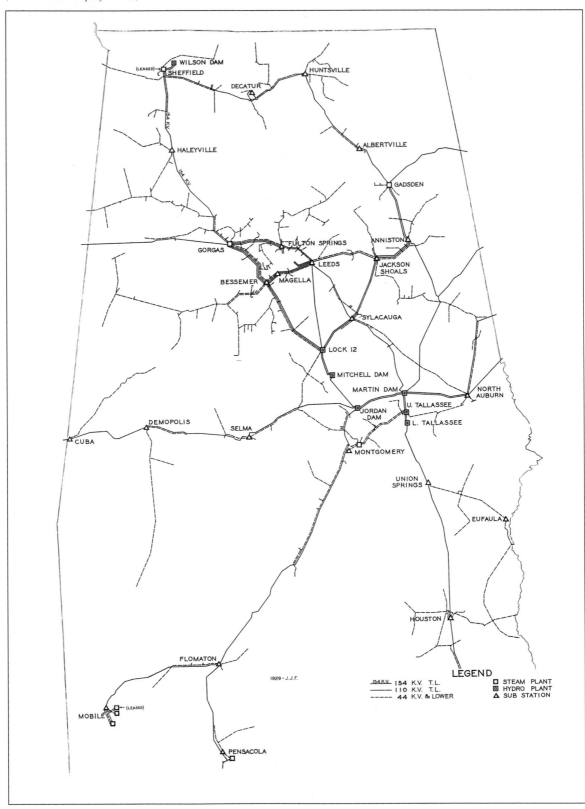

folks," they probably realized that their own circumstances surely would have been far worse if Alabama Power had not come into their communities and built those dams. And they also surely realized that had it not been for the electricity being sent out to towns, cities, and factories, the impact of the depression on all Alabama's citizens would have been far greater than it was. In fewer than two decades of frantic activity, the Alabama Power Company had harnessed the Coosa and Tallapoosa and had put those "loafing streams" to work. What might have been accomplished, had it not been for the depression, will never be known. What was accomplished before the crisis came continues to be a source of amazement.

APCO Biographies
1906-1930

(This appendix and those that follow were prepared by Bill Tharpe, Alabama Power Company archivist, from information in the Power Company archives.)

W. P. Lay

1853-1940
Steamboat captain, builder of Gadsden's first electric plant, pioneer developer of the Coosa River, founder and first president of Alabama Power Company, 1906.

James Mitchell

1866-1920
Pioneer American Hydro and Electric System developer. Executive head of Alabama Power Company from the beginning of its active work in 1912 until his death in 1920. President, APCO, 1912-13, 1915-20.

Frank S. Washburn

President, American Cyanamid Company. Organized hydro development interests on Tennessee River and was participant in Tallapoosa River Hydro Development Group. Both groups merged with Alabama Power Company in 1912. President, APCO, 1913-14.

Thomas W. Martin

1881-1964
Lawyer, businessman, scientist, historian, industrial and civic leader. Legal counsel, Tallapoosa River Hydro Development Group. Part of James Mitchell's team when hydro interests on the Coosa, Tallapoosa, and Tennessee were consolidated to form the Alabama Power Company, 1912. Executive head of Alabama Power Company 1920-63. Founder of Southern Research Institute, 1941.

Reuben A. Mitchell
Sidney Z. Mitchell

Brothers who were early electrical pioneers in electric industry in Alabama. Owned utility companies in Decatur, Huntsville, Talladega, Anniston, and elsewhere in the state. These holdings were acquired by Alabama Power Company in 1912. Directors of Alabama Power Company and other associated companies. Reuben, senior vice president, 1920-36. Jordan Dam named in honor of the Mitchells' mother, Elmira Sophia Jordan.

194

Eugene A. Yates   Alabama Power Company's first Chief Engineer, 1912–16; oversaw construction of Lock 12 (Lay) Hydro Plant, Gadsden Steam Plant, and the company's initial transmission lines and was instrumental in other early construction projects. Vice president and general manager, 1921–30.

Oscar G. Thurlow   Designing Engineer, Lock 12 (Lay) Hydro Plant and played important role in engineering of other early hydro projects. Construction manager of Dixie Construction Company, 1917–23; chief engineer, 1916–30; vice president, 1924–30.

A. C. Polk   Resident engineer during construction of Lock 12 (Lay) Dam, 1912–14. Construction manager of Dixie Construction Company, 1923–? Oversaw construction of Upper Tallassee (Yates), Tallassee Falls (Thurlow), and Cherokee Bluffs (Martin) Dams.

C. C. Davis   Coordinated shipments during construction of Warrior (Gorgas) Steam Plant. Superintendent of construction at Mitchell, Cherokee Bluffs (Martin), and Lock 18 (Jordan) Dams.

Hydroelectricity from Alabama Power

Lay Dam and Hydroelectric Generating Plant

LOCATION
>   Town: Near Clanton
>   Counties: Chilton and Coosa
>   River: Coosa
>   River miles above Mobile: 410 miles

CONSTRUCTION STARTED
>   Original: 1910
>   Redevelopment: August 31, 1964

IN-SERVICE DATE
>   Original: 1914
>   Redevelopment: February 24, 1967

DAM
>   Type: Gravity concrete
>   Length: 2,260 feet
>   Maximum height: 129.6 feet
>   Volume of concrete: 343,985 cubic yards
>   Spillway gates: 26–30 feet by 17 feet
>   Capacity of each gate: 2,823,000 gallons per minute

RESERVOIR
>   Elevation above sea level: 396 feet
>   Area: 12,000 acres
>   Shoreline: 289 miles
>   Maximum depth at dam: 88 feet
>   Area of watershed draining into reservoir: 9,087 square miles
>   Power storage: 27,570,000 kilowatt-hours

DEDICATION
>   November 23, 1929

POWERHOUSE
>   Length: 376 feet
>   Height: 106 feet

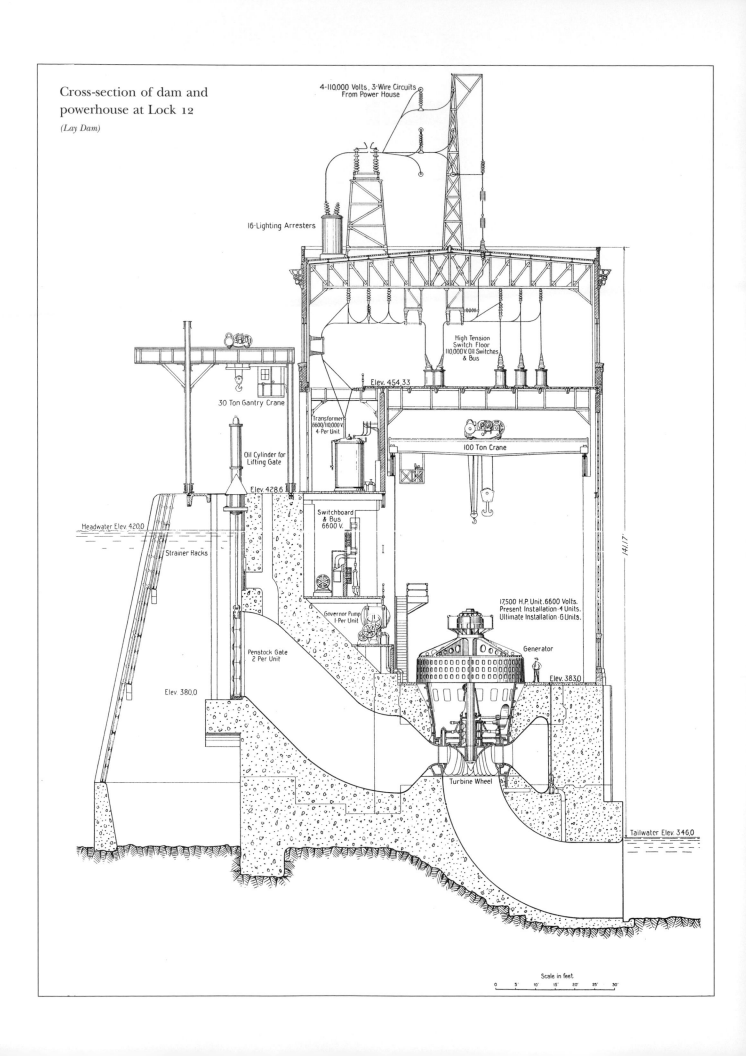

Cross-section of dam and
powerhouse at Lock 12

(Lay Dam)

4-110,000 Volts, 3-Wire Circuits
From Power House

16-Lighting Arresters

High Tension
Switch Floor
110,000 V. Oil Switches
& Bus

Elev. 454.33

30 Ton Gantry Crane

Transformer
6600/110,000 V.
4-Per Unit

100 Ton Crane

Oil Cylinder for
Lifting Gate

Elev. 428.6

Headwater Elev. 420.0

Switchboard
& Bus
6600 V.

Strainer Racks

17,500 H.P. Unit. 6600 Volts.
Present Installation-4 Units.
Ultimate Installation-6 Units.

Governor Pump
1-Per Unit

Generator

Penstock Gate
2 Per Unit

Elev. 383.0

Elev. 380.0

Turbine Wheel

141.17'

Tailwater Elev. 346.0

Scale in feet.
0   5'   10'   15'   20'   25'   30'

Width: 74 feet
Crane capacity: 110 tons

HYDRAULIC TURBINES
Number: 6
Type: Mixed Flow
Manufacturer: Allis Chalmers
Horsepower at 81 feet head: 40,000 each
Water discharge per turbine: 2,477,000 gallons per minute
Speed: 128.6 revolutions per minute
Diameter of water wheel: 14 feet, 7 inches
Weight of water wheel: 63,900 pounds

ALTERNATING CURRENT GENERATORS
Number: 6
Manufacturer: General Electric
Rating: 29,500 kilowatts each
Voltage: 11,500 volts
Speed: 128.6 revolutions per minute
Diameter of rotor: 22 feet, 8 inches
Weight of rotor: 253,000 pounds
Estimated average annual output: 581,400,000 kilowatt-hours

POWER TRANSFORMERS
Number: 2
Voltage (low side): 11,500 volts
Voltage (high side): 115,000 volts
Rating: 105,000 kilovolt-amperes each

Lay Dam was named after Alabama Power Company's first president, Capt. William Patrick Lay. He organized the company on December 4, 1906. Then he got authorization from Congress to construct the company's first dam and electric generating plant on the Coosa River. First known as the Lock 12 Dam, the facility was later named in recognition of Captain Lay's service to the company and to the public.

Lay Dam was redeveloped during the 1960s as part of the Coosa River Project. The project included the construction of H. Neely Henry, Weiss, Logan Martin, and Bouldin dams.

Mitchell Dam and Hydroelectric Generating Plant

LOCATION
Town: Near Verbena
Counties: Chilton and Coosa
River: Coosa
River miles above Mobile: 397 miles

198 Construction started
          Units 1–3: July 1, 1921
          Unit 4: June 1948
          Units 5–7: October 26, 1977

In-service date
          Units 1–3: August 15, 1923
          Unit 4: November 9, 1949
          Units 5–7: April 1, 1985

Units 1–3 were retired from service
          May 1, 1985

Dam

          Type: Gravity concrete
          Length: 1,277 feet
          Maximum height: 106 feet
          Volume of concrete: 340,569 cubic yards
          Spillway gates: 22–30 feet by 15 feet, 3–30 feet by 25 feet
          Capacity of each gate: Small—2,747,000 gallons per minute
               Large—6,250,000 gallons per minute

Reservoir
          Elevation above sea level: 312 feet
          Area: 5,850 acres
          Shoreline: 147 miles
          Length: 14 miles
          Maximum depth of dam: 90 feet
          Area of watershed draining into reservoir: 9,827 square miles
          Power storage: 8,000,000 kilowatt-hours

Powerhouses
          (Unit 4)
          Length: 449 feet
          Height: 141 feet
          Width: 83 feet
          Crane capacity: 125 tons

          (Units 5–7)
          Length: 300 feet
          Height: 163 feet
          Width: 90 feet
          Crane capacity: 250 tons

Hydraulic turbines
          (Unit 4)
          Number: 1

Type: Francis
Manufacturer: Allis Chalmers
Horsepower at 70 feet head: 29,000
Water discharge: 2,150,000 gallons per minute
Speed: 90 revolutions per minute
Diameter of water wheel: 14 feet, 10 inches
Weight of water wheel: 80,500 pounds

(Units 5–7)
Number: 3
Type: Fixed blade propeller, 5 blades
Manufacturer: Allis Chalmers
Horsepower at 57.4 feet head: 69,000 each
Water discharge per turbine: 4,694,000 gallons per minute
Speed: 85.7 revolutions per minute
Diameter of water wheel: 22 feet, 2 inches
Weight of water wheel: 140,300 pounds

ALTERNATING CURRENT GENERATORS
(Unit 4)
Number: 1
Manufacturer: General Electric
Rating: 20,000 kilowatts
Voltage: 6,600 volts
Speed: 90 revolutions per minute
Diameter of rotor: 26 feet, 4 inches
Weight of rotor: 364,000 pounds

(Units 5–7)
Number: 3
Manufacturer: Siemans-Allis
Rating: 50,000 kilowatts each
Voltage: 13,800 volts
Speed: 85.7 revolutions per minute
Diameter of rotor: 31 feet, 5 inches
Weight of rotor: 46,000 pounds
Estimated average annual output: 700,000,000 kilowatt-hours

POWER TRANSFORMERS
(Unit 4)
Voltage (low side): 6,600 volts
Voltage (high side): 115,000 volts
Rating: 25,000 kilovolt-amperes

(Units 5–7)
Voltage (low side): 13,800 volts
Voltage (high side): 115,000 volts
Rating: 185,920 kilovolt-amperes

Mitchell Dam is the second oldest of fourteen Alabama Power Company hydroelectric plants. The dam was named for James Mitchell, Alabama Power president from 1912 to 1920.

Because of financial difficulty construction of units 5–7 was delayed from December 27, 1978, until April 12, 1981. A unique feature of the new powerhouse is a 1,140-foot floating trash boom that deflects trash from the powerhouse intakes. A fishing facility located below the dam is open year-round to the public. Parking, restrooms, picnic tables, and an overlook are also available for public use.

Martin Dam and Hydroelectric Generating Plant

LOCATION
> Town: Near Dadeville
> Counties: Elmore and Tallapoosa
> River: Tallapoosa
> River miles from Mobile: 420 miles

CONSTRUCTION STARTED
> Units 1, 2, and 3: July 24, 1923

IN-SERVICE DATE
> December 31, 1926

UNIT 4 INSTALLED
> 1952

DAM
> Type: Gravity concrete
> Length: 2,000 feet
> Maximum height: 168 feet
> Volume of concrete: 431,000 cubic yards
> Spillway gates: 20–30 feet by 16 feet
> Capacity of each gate: 3,016,000 gallons per minute

RESERVOIR
> Elevation above sea level: 490 feet
> Area: 40,000 acres
> Shoreline: 700 miles
> Length: 31 miles
> Maximum depth at dam: 155 feet
> Area of watershed draining into reservoir: 3,000 square miles
> Power storage: 301,700,000 kilowatt-hours (Includes Yates and Thurlow dams)

POWERHOUSE
> Length: 307.9 feet
> Height: 99 feet

Width: 58 feet
Crane capacity: 200 tons

HYDRAULIC TURBINES
Number: 4
Type: Francis
Manufacturers: I. P. Morris, Allis Chalmers, and S. M. Smith
Horsepower at 145 feet head: *78,000
Water discharge: *2,744,000 gallons per minute
Speed: *112.5 revolutions per minute
Diameter of water wheel: *15 feet
Weight of water wheel: *118,000 pounds
*Largest unit

ALTERNATING CURRENT GENERATORS
Number: 4
Manufacturer: General Electric
Rating: *55,200 kilowatts
Voltage: 12,000 volts
Speed: *112.5 revolutions per minute
Diameter of rotor: *27 feet, 1 inch
Weight of rotor: *484,150 pounds
Estimated average annual output: 336,000,000 kilowatt-hours
*Largest unit

POWER TRANSFORMER
Number: 4
Voltage (low side): 12,000 volts
Voltage (high side): 115,000 volts
Rating (largest unit): 66,000 kilovolt-amperes

Martin Dam was the first of four dams constructed on the Tallapoosa River. When it was built, the dam created the world's largest artificial body of water.

First known as Cherokee Bluffs, the dam was dedicated in 1936 in honor of Thomas Martin, president of Alabama Power Company from 1920 to 1949 and chief executive officer from 1949 to 1963. Martin was instrumental in the development of Alabama Power and a pioneer in the development of the electric system throughout Alabama and the Southeast.

Jordan Dam and Hydroelectric Generating Plant

LOCATION
Town: Near Wetumpka
County: Elmore
River: Coosa
River miles above Mobile: 378 miles

CONSTRUCTION STARTED
June 15, 1926

IN-SERVICE DATE
December 31, 1928

DAM

Type: Gravity concrete
Length: 2,066 feet
Maximum height: 125 feet
Volume of concrete: 389,000 cubic yards
Spillway gates: 17–30 feet by 18 feet, 18–34 feet by 8 feet
Capacity of each gate: Large—3,734,000 gallons per minute
   Small—956,000 gallons per minute

RESERVOIR

Elevation above sea level: 252 feet
Area: 6,800 acres
Shoreline: 118 miles
Length: 18.4 miles
Maximum depth at dam: 110 feet
Area of watershed draining into reservoir: 10,165 square miles

POWERHOUSE

Length: 300 feet
Height: 105 feet
Width: 62 feet
Crane capacity: 150 tons

HYDRAULIC TURBINES

Number: 4
Type: Francis
Manufacturers: S. M. Smith
Horsepower at 90 feet head: 36,000 each
Water discharge per turbine: 2,089,000 gallons per minute
Speed: 100 revolutions per minute
Diameter of water wheel: 14 feet 10 inches
Weight of water wheel: 120,000 pounds

ALTERNATING CURRENT GENERATORS

Number: 4
Manufacturer: Westinghouse
Rating: 25,000 kilowatts each
Voltage: 12,000 volts
Speed: 100 revolutions per minute
Diameter of rotor: 25 feet 4 inches
Weight of rotor: 310,000 pounds

Estimated average annual output: 148,543,000 kilowatt-hours   **203**

POWER TRANSFORMERS
Number: 2
Voltage (low side): 12,000 volts
Voltage (high side): 115,000 volts
Rating: 60,000 kilovolt-amperes each

When Jordan Dam was built, it was the largest power project undertaken by private capital in the South.

Jordan Dam was named in honor of the brothers Reuben and Sidney Mitchell, who played an important part in the early development of the company. Since there was already a Mitchell Dam, named in honor of Alabama Power Company President James Mitchell (no relation), the brothers chose to give the dam their mother's maiden name.

Yates Dam and Hydroelectric Generating Plant

LOCATION
Town: Near Tallassee
Counties: Elmore and Tallapoosa
River: Tallapoosa
River miles above Mobile: 412 miles

CONSTRUCTION STARTED
January 18, 1927

IN-SERVICE DATE
July 1, 1928

DAM
Type: Gravity concrete
Length: 1,261 feet
Maximum height: 87 feet
Volume of concrete: 125,000 cubic yards

RESERVOIR
Elevation above sea level: 344 feet
Area: 2,000 acres
Shoreline: 40 miles
Length: 7.9 miles
Maximum depth at dam: 46.5 feet
Area of watershed draining into reservoir: 3,250 square miles

POWERHOUSE
Length: 203 feet
Height: 77 feet

Width: 59 feet
Crane capacity: 150 tons

HYDRAULIC TURBINES
Number: 2
Type: Francis
Manufacturer: I. P. Morris
Horsepower at 55 feet head: 25,000 each
Water discharge per turbine: 1,997,000 gallons per minute
Speed: 80 revolutions per minute
Diameter of water wheel: 16 feet, 3 inches
Weight of water wheel: 144,000 pounds

ALTERNATING CURRENT GENERATORS
Number: 2
Manufacturer: General Electric
Rating: 16,000 kilowatts each
Voltage: 6,900 volts
Speed: 80 revolutions per minute
Diameter or rotor: 25 feet, 9 inches
Weight of rotor: 238,000 pounds
Estimated average annual output: 136,000,000 kilowatt-hours

POWER TRANSFORMER
Voltage (low side): 6,900 volts
Voltage (high side): 44,000/115,000 volts
Rating (largest unit): 40,000 kilovolt-amperes

Yates Dam was named for Eugene A. Yates in recognition of his outstanding contributions to Alabama Power Company and to its customers as an engineer and administrator. Yates joined Alabama Power in 1912 as chief engineer and was responsible for completing Lay Dam, the company's first hydroelectric plant.

Though Yates Dam is modest in size and capacity when compared with other Alabama Power plants, its location gives it special significance. The dam stands on the site of Alabama's first hydroelectric plant, which began delivering electricity over a twenty-five-mile transmission line to Montgomery in 1912. At the dedication ceremony, Alabama Power Company President Thomas Martin said that the location of Yates Dam "should ever be recognized as a symbol of progress, of foresight, and of scientific and economic research."

Thurlow Dam and Hydroelectric Generating Plant

LOCATION
Town: Tallassee
Counties: Elmore and Tallapoosa
River: Tallapoosa
River miles above Mobile: 409 miles

CONSTRUCTION STARTED
>April 5, 1928

IN-SERVICE DATE
>December 31, 1930

DAM
>Type: Gravity concrete
>Length: 1,846 feet
>Maximum height: 62 feet
>Volume of concrete: 150,000 cubic yards

RESERVOIR
>Elevation above sea level: 288.8 feet
>Area: 574 acres
>Shoreline: 6 miles
>Length: 3 miles
>Maximum depth at dam: 54 feet
>Area of watershed draining into reservoir: 3,300 square miles

POWERHOUSE
>Length: 185 feet
>Height: 69 feet
>Width: 54 feet
>Crane capacity: 150 tons

HYDRAULIC TURBINES
>Number: 3
>Type: Francis
>Manufacturers: I. P. Morris and S. M. Smith
>Horsepower at 88 feet head: *36,000
>Water discharge: *1,706,000 gallons per minute
>Speed: *100 revolutions per minute
>Diameter of water wheel: *14 feet, 10 inches
>Weight of water wheel: *90,000 pounds
>*Largest unit

ALTERNATING CURRENT GENERATORS
>Number: 3
>Manufacturer: Westinghouse and General Electric
>Rating: *25,000 kilowatts
>Voltage: 13,800 volts
>Speed: *100 revolutions per minute
>Diameter of rotor: *25 feet, 3 inches
>Weight of rotor: *213,300 pounds
>Estimated average annual output: 245,154,000 kilowatt-hours
>*Largest unit

206

POWER TRANSFORMER
Voltage (low side): 13,800 volts
Voltage (high side): 115,000 volts
Rating (largest unit): 70,000 kilovolt-amperes

Thurlow Dam was built at the site of an early nineteenth-century textile mill that was used during the Civil War as a uniform and ammunition plant. The dam was named in honor of Oscar G. Thurlow, a chief engineer, vice president, and director of Alabama Power Company.

In 1928, Thurlow was awarded the Howard N. Potts Medal by the Franklin Institute in Philadelphia for his work in science and the mechanical arts. He invented the Thurlow Backwater Suppressor, which was first installed in Mitchell Dam on the Coosa River.

Notes

Introduction: In the Beginning

1. There have been a number of studies of the early years of the Alabama Power Company and William Patrick Lay. The most complete is John R. Hornaday, *Soldiers of Progress and Industry* (New York, 1930), see 34–39. To place Lay within the context of river development see Harvey H. Jackson, *Rivers of History: Life on the Coosa, Tallapoosa, Cahaba, and Alabama* (Tuscaloosa, Ala., 1995), chapter 13. See also Hughes Reynolds, *The Coosa River Valley from De Soto to Hydroelectric Power* (Cynthiana, Ky., 1944), 247–49.

2. The early years of the Alabama Power Company after Thomas Martin and James Mitchell became involved are covered in Thomas W. Martin, *Forty Years of Alabama Power Company* (New York, 1952), 5–13. See also Thomas W. Martin, *The Story of Electricity in Alabama since the Turn of the Century* (Birmingham, 1952), 21–28; and William M. Murray, Jr., *Thomas W. Martin: A Biography* (Birmingham, 1978), 28–29.

3. Martin, *Forty Years of Alabama Power Company*, 12–18; Murray, *Thomas W. Martin*, 80.

4. Martin, *Forty Years of Alabama Power Company*, 15–16; Murray, *Thomas W. Martin*, 30–34.

5. Martin, *Forty Years of Alabama Power Company*, 17–19; Murray, *Thomas W. Martin*, 36–39.

6. Jackson, *Rivers of History*, 178–79. Murray, *Thomas W. Martin*, 37–38.

7. Murray, *Thomas W. Martin*, 37–38. E. L. Sayers and A. C. Polk, *The Lock 12 Development of the Alabama Power Company, Coosa River, Alabama*, reprinted from the *Transactions* of the American Society of Civil Engineers, 77 (1915), 1409–1581.

8. Although books by Martin, Murray, and Hornaday present the history of the company from the perspective of management, the construction reports and letters in the files in the Alabama Power Company Archives give all points of view.

9. Jackson, *Rivers of History*, 176–78. For additional background on hydroelectric development in the region and Alabama Power Company's place in it see Martin, *The Story of Electricity in Alabama*, 1–41.

10. Jackson, *Rivers of History*, chapter 13, covers conditions in the region where the dams were built. Sayers and Polk, *Lock 12 Development*, 1529–1534, discusses laborers and the company's needs.

11. Stuart D. Brandes, *American Welfare Capitalism* (Chicago, 1976), 1–9. See also Gerald Zahavi, *Workers, Managers, and Welfare Capitalism: The Shoeworkers and Tanners of Endicott Johnson, 1890–1950* (Urbana, 1988).

12. Alabama Power's personnel policies were surely influenced by those instituted by Birmingham companies, especially Tennessee, Coal, Iron and Railroad Company and Sloss Furnaces. See Marlene Hunt Rikard, "An Experiment in Welfare Capitalism: The Health Care Services of the Tennessee, Coal, Iron and Railroad Company" (Ph.D. diss.,

University of Alabama, 1983); and W. David Lewis, *Sloss Furnaces and the Rise of the Birmingham District: An Industrial Epic* (Tuscaloosa, Ala., 1994), esp. pp. 474–511. For general social and economic conditions in Alabama during this period see William Warren Rogers, Robert David Ward, Leah Rawls Atkins, and Wayne Flynt, *Alabama: The History of a Deep South State* (Tuscaloosa, Ala., 1994).

13. *Clanton Union Banner,* March 26, 1914.

14. For an excellent analysis of the relationship between workers and management in other Alabama industrial settings see Henry M. McKiven, Jr., *Iron & Steel: Class, Race, and Community in Birmingham, Alabama, 1875–1920* (Chapel Hill, N.C., 1995).

15. Insight into the history of black-white labor relations can be found in Jack Temple Kirby, "Negotiators/Nonnegotiators," in *The Countercultural South* (Athens, Ga., 1995), 8–32. See also McKiven, *Iron & Steel.*

Chapter One. A Dam at Lock 12

1. *Clanton Union Banner,* November 24, 1921; Anita Sanders Bosley, "Looking Back: Belle Hendrix, Growing Up with Lay Dam," *Powergrams* (February 1990), 7.

2. *Clanton Union Banner,* November 24, 1921; interview with L. B. Crouch, April 23, 1995, Tallassee, Alabama. This same story was told by L. B. "Gip" O'Daniel about Martin Dam. Interview on March 10, 1995, at Tallassee, Alabama. Copies of the tapes and transcripts of the interviews that were conducted by the author are in the Alabama Power Company Archives, Birmingham, Alabama. A second copy is in the Houston Cole Library, Jacksonville State University, Jacksonville, Alabama.

3. Jackson, *Rivers of History,* 144–46, 173–74.

4. Bosley, "Belle Hendrix," 7–8.

5. Ibid.

6. *Clanton Press,* March 16, 1911; *Birmingham News,* February 14, 1912.

7. [A. C. Polk, Resident Engineer], *History of Construction of Lock 12: Dam and Foundations: Coosa River,* vol. 1, typescript copy in the Alabama Power Company Archives, Birmingham, Alabama, 8–9; Sayers and Polk, *The Lock 12 Development,* 1455–56, 1545. Information on MacArthur Brothers is sketchy, but Mitchell's contacts with other hydroelectric developers would have kept him aware of the best contractors in the field. See Martin, *The Story of Electricity in Alabama,* 25–34, and S. A. Mitchell, *S. Z. Mitchell and the Electrical Industry* (New York, 1960).

8. *History of Construction of Lock 12 Dam,* 155–56; *Lock 12 Development,* 1454.

9. *History of Construction of Lock 12 Dam,* 9–10, 156–58.

10. *History of Construction of Lock 12 Dam,* 10–14, 22, 158–61; *Lock 12 Development,* 1455–59, 1531.

11. *Lock 12 Development,* 1478, 1536–37; Thomas W. Martin to Dr. H. L. Castleman (Talladega County health officer), December 27, 1912, Lay Dam Files, Alabama Power Company Archives.

12. *History of Construction of Lock 12 Dam,* 22–28; *Lock 12 Development,* 1460; J. A. Le Prince, *A Study of Impounded Waters on the Coosa River in Shelby, Chilton, Talladega and Coosa Counties Alabama* (Birmingham, n.d. [1915]), 3–4; Dr. W. H. Sanders to Thomas Martin, March 8, 1913, Lay Dam Files, Alabama Power Company Archives.

13. *Birmingham Age-Herald,* January 5, 1913; *Lock 12 Development,* 1461–1529, describes building the dam.

14. *Birmingham News*, January 4, 1913; *Birmingham Age-Herald,* January 5, 1913.

15. *History of Construction of Lock 12 Dam,* 162; *Lock 12 Development,* 1531–32.

16. *History of Construction of Lock 12 Dam,* 162–63; *Lock 12 Development,* 1531.

17. *Lock 12 Development,* 1532.

18. *Montgomery Advertiser,* June 5, 1913; *Lock 12 Development,* 1465–66, 1529–34; *History of Construction of Lock 12 Dam,* 161–73, appendix.

19. *History of Construction of Lock 12 Dam,* 163–64.

20. Ibid., 29–30.

21. Ibid., 30–33. For additional information on workers' camps in the Birmingham District, see Marlene Hunt Rikard, "An Experiment in Welfare Capitalism: The Health Care Services of Tennessee Coal, Iron, and Railroad Company" (Ph.D. diss., University of Alabama, 1983), 91–130; and McKiven, *Iron & Steel,* 55–76. See also Lewis, *Sloss Furnaces.*

22. *History of Construction of Lock 12 Dam,* 29–35; *Birmingham Age-Herald,* January 5, 1913.

23. *History of Construction of Lock 12 Dam,* 31.

24. *History of Construction of Lock 12 Dam,* 31. See also Kirby, "Negotiators/Nonnegotiators," 8–32.

25. *Lock 12 Development,* 1466–67; *History of Construction of Lock 12 Dam,* 32. See also Kirby, "Negotiators/Nonnegotiators."

26. *History of Construction of Lock 12 Dam,* 32; *Birmingham Age-Herald,* January 5, 1913.

27. *History of Construction of Lock 12 Dam,* appendix, 1–3, 5–6.

28. *Lock 12 Development,* 1466.

29. *History of Construction of Lock 12 Dam,* appendix, 1–8. F. H. Chamberlain (general manager) to Frank S. Washburn (president), May 29, 1914, Lay Dam Files, Alabama Power Company Archives; *Birmingham News,* March 24, 1914.

30. *History of Construction of Lock 12 Dam,* 3–4, appendix, 1–8

31. *History of Construction of Lock 12 Dam,* appendix, 6–11.

32. *History of Construction of Lock 12 Dam,* appendix, 8–11.

33. *Lock 12 Development,* 1461, 1482–94.

34. *Lock 12 Development,* 1524–29.

35. *Lock 12 Development,* 1535.

36. *Lock 12 Development,* 1462.

37. Both [Polk], *History of Construction of Lock 12 Dam,* and Polk and Sayers, *Lock 12 Development,* contain numerous recommendations for improvements that can be made. See also *History of Construction of Lock 12 Dam,* 35–36; *Lock 12 Development,* 1464–66. For a thorough discussion of workers' medical care in the Birmingham steel industry, which probably influenced Power Company officials, see Rikard, "An Experiment," 131–268.

38. *History of Construction of Lock 12 Dam,* 158–60; *Lock 12 Development,* 1460, 1530–31.

39. *History of Construction of Lock 12 Dam,* 30–31, 167–68, 173–74; *Lock 12 Development,* 1462–65.

Chapter Two. Almost Done in by a Mosquito

1. *The Chronicle* (Rockford, Ala.), January 10, 1913. For statewide comments see *Birmingham News,* February 14, 1912; *Montgomery Advertiser,* June 5, 1913.

2. Dr. H. L. Castleman to Alabama Power Company, August 10, 1913, Lay Dam Files, Alabama Power Company Archives.

3. Thomas Martin to Dr. H. L. Castleman, December 27, 1913; E. A. Yates, "Clearing Lock 12 Reservoir," typescript report, pages not numbered. Both are in the Lay Dam Files, Alabama Power Company Archives; *Lock 12 Development*, 1536–37.

4. Prof. Edgar B. Kay to Dr. W. H. Sanders, January 18, 1913, copy in "Clearing Lock 12 Reservoir."

5. "Clearing Lock 12 Reservoir."

6. Ibid.; Hatcher and Smith (law firm) to Thomas Martin, February 10, 1913; Dr. W. H. Sanders to Thomas W. Martin, March 8, 1913, both in the Lay Dam Files, Alabama Power Company Archives.

7. T. L. Stewart to Thomas W. Martin, August 30, 1913; T. L. Stewart to Edward L. Sayers, August 30, 1913; W. S. Hulse to T. L. Stewart, August 31, 1913; Thomas Martin to E. A. Yates, September 2, 1913, all in the Lay Dam Files, Alabama Power Company Archives.

8. Thomas Martin to E. A. Yates, September 2, 1913; J. K. Dixon to Thomas W. Martin, September 4, 1913; E. E. Watters to Marion H. Simms, September 2, 1913; Edward Sayers to Thomas W. Martin, September 6, 1913, all the Lay Dam Files, Alabama Power Company Archives.

9. R. R. Burkhalter to E. L. Sayers, [September, 1913]; J. K. Dixon to Thomas W. Martin, October 18, 1913; E. A. Yates to Thomas Martin, October 20, 1913; Thomas W. Martin to J. K. Dixon, October 20, 1913; E. A. Yates to Thomas Martin, October 20, 1913, all in Lay Dam Files, Alabama Power Company Archives.

10. J. K. Dixon to Thomas W. Martin, October 18, 1913; J. K. Dixon to Thomas Martin, October 22, 1913; E. A. Yates to Thomas Martin, October 20, 1913; P. W. Welch to Dr. W. H. Sanders, October 22, 1913; Thomas Martin to E. A. Yates, November 6, 1913, all in the Lay Dam Files, Alabama Power Company Archives.

11. P. W. Welch to Dr. W. H. Sanders, October 22, 1913; T. L. Stewart to Thomas Martin, November 8, 1913, and December 22, 1913, all in the Lay Dam Files, Alabama Power Company Archives.

12. T. L. Stewart to Thomas Martin, December 22, 1913, Lay Dam Files, Alabama Power Company Archives.

13. Ibid.

14. B. R. Powell to Thomas W. Martin, January 5, 1914; Dr. W. H. Sanders to Thomas W. Martin, February 13, 1914, both in Lay Dam Files, Alabama Power Company Archives.

15. Thomas W. Martin to Dr. W. H. Sanders, February 14, 1914; Thomas W. Martin to E. A. Yates, February 14, 1914; Thomas W. Martin to Frank S. Washburn, February 14, 1914, all in Lay Dam Files, Alabama Power Company Archives.

16. W. E. Mitchell to F. H. Chamberlain, May 12, 1914; F. H. Chamberlain to Frank S. Washburn, September 23, 1914, both in Lay Dam Files, Alabama Power Company Archives. For more information on efforts to control malaria in industrial camps, see Marlene Hunt Rikard, "George Gordon Crawford: Man of the New South," *Alabama Review* 31 (July 1978): 165–70.

17. W. E. Mitchell to F. H. Chamberlain, May 12, 1914, Lay Dam Files, Alabama Power Company Archives; *Clanton Union Banner,* May 14, 1914; Bosley, "Looking Back:

Belle Hendrix," *Powergrams* (February 1990), 8–9. Labor turnover because of sickness was a serious problem in most industrial settings. See Rikard, "George Gordon Crawford," 165–70.

18. Thomas W. Martin to W. E. Mitchell, September 18, 1914; Thomas W. Martin to Frank S. Washburn, September 21, 1914, both in Lay Dam Files, Alabama Power Company Archives.

19. Thomas W. Martin to W. E. Mitchell, September 18, September 21, September 28, 1914, all in Lay Dam Files, Alabama Power Company Archives.

20. V. J. Gregg (?) to Dr. W. H. Sanders, September 25, October 14, 1914, both letters in Lay Dam Files, Alabama Power Company Archives.

21. S. D. Motley to Dr. W. H. Sanders, October 14, 1914; O. G. Thurlow to F. H. Chamberlain, September 30, 1914; Le Prince, *Impounded Waters*, 19–36.

22. Jack Kytle, "I'm Allus Hongry," in James Seay Brown, Jr., ed., *Up before Daylight: Life Histories from the Alabama Writers' Project, 1938–39* (University, Ala., 1982), 122, 125–26; Le Prince, *Impounded Waters*, 19–36.

23. Drs. J. T. Hunter, C. J. S. Peterson, C. K. Maxwell, and Julius Jones to Dr. W. H. Sanders, October 29, 1914, Lay Dam Files, Alabama Power Company Archives.

24. Mosquito survey, November and December 1914; Frank S. Washburn to Thomas W. Martin, September 28, 1914, both in the Lay Dam Files, Alabama Power Company Archives. See also *Birmingham News*, February 14, 1915.

25. P. W. Welch to Col. W. C. Gorgas, October 6, 1914; W. C. Gorgas to P. W. Welch, October 19, 1914, both in Lay Dam Files, Alabama Power Company; Martin, *Forty Years of Alabama Power Company*, 20; Murray, *Thomas W. Martin*, 40–43; *Birmingham News*, February 14, 1915.

26. F. H. Chamberlain to W. E. Mitchell, February 8, 1915; O. G. Thurlow to Dr. H. R. Carter, February 17, 1915, both in Lay Dam Files, Alabama Power Company.

27. O. G. Thurlow to Dr. H. R. Carter, February 17, 1915, Lay Dam Files, Alabama Power Company; Martin, *Forty Years*, 20–21; Murray, *Thomas W. Martin*, 42–43; *Birmingham News*, February 14, 1915.

28. *Birmingham News*, February 14, 1915.

29. Kytle, "I'm Allus Hongry," 126.

30. Martin, *Forty Years*, 19–20; Murray, *Thomas W. Martin*, 45–51.

Chapter Three. Gathering Streams from Waste

1. O. G. Thurlow to Dr. H. R. Carter, February 17, 1915, Lay Dam Files, Alabama Power Company Archives.

2. O. G. Thurlow to F. H. Chamberlain, April 27, 1915; O. G. Thurlow to W. E. Mitchell, April 29, 1915, both in Lay Dam Files, Alabama Power Company Archives.

3. O. G. Thurlow to W. E. Mitchell, April 29, 1915, Lay Dam Files, Alabama Power Company Archives.

4. Le Prince, *Impounded Waters*, 14–16; D. L. Van Dine to O. C. Thurlow, December 17, 1914; O. C. Thurlow to Dr. R. H. von Ezdorf, January 5, 1915; Dr. R. H. von Ezdorf to O. G. Thurlow, January 8, 1915; O. G. Thurlow to Dr. R. H. Carter, January 20, 1915; O. G. Thurlow to W. E. Mitchell, April 29, 1915, all in Lay Dam Files, Alabama Power Company Archives.

5. John H. Wallace, Jr., to Thomas W. Martin, June 4, 1915; John H. Wallace, Jr., to Thomas W. Martin, June 18, 1915; E. E. Watters to John H. Wallace, June 16, 1915; Thomas W. Martin to John H. Wallace, Jr., June 23, 1915, all in Lay Dam Files, Alabama Power Company Archives.

6. Martin, *Forty Years*, 10, 18; Murray, *Thomas W. Martin*, 46–47; Jackson, *Rivers of History*, 180–81.

7. *Clanton Union Banner*, April 12, 1917.

8. Martin, *Forty Years*, 24; Murray, *Thomas W. Martin*, 54–56.

9. Martin, *Forty Years*, 26–27; "Alabama Power Company Dedication of Mitchell Dam to the Service of the Public and in Memory of Mr. James Mitchell, December 19, 1921," including addresses by Thomas W. Martin and W. E. Mitchell, in Mitchell Dam File, Alabama Power Company Archives.

10. "Mitchell Dam Project," *Powergrams* (October 1921), 12; W. N. Walmsley to James Mitchell, September 8, 1917, and an interview with Mr. J. E. Ladd, June 9, 1974, both in the Dixie Construction File, Alabama Power Company Archives.

11. S. S. Simpson, "The Function of the Dixie Construction Company and the Men Who Make It Function," *Powergrams* (March 1925), 12–13; "Resolution adopted at meeting of Board of Director of Alabama Power Company," October 16, 1917, Dixie Construction File, Alabama Power Company Archives; "Before the Federal Power Commission in the Matter of Alabama Power Company, Jordan Dam, Project No. 618, Determination of Actual Legitimate Original Cost as of December 31, 1929, Brief on Behalf of Licensee, June 10, 1940" (Birmingham Printing Company), 84–85. This brief contains testimony by A. C. Polk on the origins of Dixie Construction.

12. *Birmingham Age-Herald, Montgomery Advertiser, Gadsden Evening Journal,* and *Gadsden Times News* published the announcement on November 5, 1920. See also Bosley, "Looking Back: Belle Hendrix," 8.

13. *Clanton Union Banner*, January 8, June 10, June 22, June 24, 1920.

14. *Clanton Union Banner*, November 11, 1920.

15. "Mitchell Dam Project," *Powergrams* (October 1921), 12–13; "Ripples from Lock 12," *Powergrams* (December 1920), 9; *Gadsden Evening Journal*, January 7, 1921.

16. *Birmingham News*, June 28, 1921; *Clanton Union Banner*, July 7, 1921; *Montgomery Journal*, October 20, 1921; "Thomas W. Martin, the Man, and His Work," *Alabama Sportsman* (August-September 1929), 6–7. See also B. P. Powell to O. G. Thurlow, August 19, 1920; "Duncan's Riffle Reservoir Estimate," Mitchell Dam File, Alabama Power Company Archives.

17. W. M. Walmsley to Thomas W. Martin, July 1, 1921, Mitchell Dam Files, Alabama Power Company Archives; Le Prince, *Impounded Waters*, 4–5; *Montgomery Advertiser*, September 18, 1921.

18. Superintendent of line construction to O. G. Thurlow, December 20, 1920; W. M. Walmsley to Walter M. Hood, November 17, 1921, both in Mitchell Dam File, Alabama Power Company Archives.

19. *Clanton Union Banner*, June 9, 1921.

20. Ibid.

21. *Clanton Union Banner*, June 23, 1921.

22. Ibid.

23. See "Ripples from Lock 12," *Powergrams* (December 1920), 20, (February 1921), 6, and (March 1921), 9.

24. A. M. Kennedy, "Mitchell Dam," *Powergrams* (September 1921), 4; *Montgomery Advertiser,* September 18, 1921.

25. "Mitchell Dam," *Powergrams* (September 1921), 4; *Birmingham News,* September 12, 1921.

26. *Powergrams* (November 1921), 2, 24, (February 1922), 18; *Birmingham Age-Herald,* October 15, 1921; *Montgomery Advertiser,* September 18, 1921.

27. *Powergrams* (February 1922), 18, (June 1922), 9; *Montgomery Journal,* December 18, 1921.

28. *Powergrams* (June 1922), 9, 12, (August 1922), 19; *Birmingham Post,* December 20, 1921; *Montgomery Advertiser,* December 24, 1922; *Montgomery Journal,* October 30, 1921; *Clanton Union Banner,* November 24, 1921; interview with L. B. Crouch, April 23, 1995, Tallassee, Alabama.

29. *Clanton Union Banner,* December 22, 1921; *Powergrams* (January 1922) issue is devoted to the dedication of Mitchell Dam. See also the program and other material from the ceremony in the Mitchell Dam File, Alabama Power Company Archives.

Chapter Four. Building Mitchell Dam

1. A complete run of *Powergrams* is in the archives of the Alabama Power Company in Birmingham.

2. *Powergrams* (October 1921), 12–13, (August 1922), 1–2.

3. *Clanton Union Banner,* July 14, 1921 (two articles).

4. *Powergrams* (October 1921), 13, 23, (December 1923), 12; Fred Mayfield to Harvey H. Jackson III, April 24, 1995, in possession of the author.

5. *Montgomery Advertiser,* September 18, 1921; *Birmingham News,* September 18, 1921; *Powergrams* (October 1921), 23.

6. *Powergrams* (October 1921), 13, (December 1923), 12–13.

7. *Powergrams* (October 1921), 23, (November 1923), 14–15, (December 1923), 12–13. Interview with Geraldine Hollis (Mrs. Harold Lawrence), April 26, 1995, Clanton, Alabama.

8. *Powergrams* (October 1921), 23, (December 1923), 13; *Montgomery Advertiser,* September 18, 1921.

9. *Birmingham Age-Herald,* October 15, 1921. See also *Birmingham News,* July 17, September 18, 1921.

10. *Birmingham Age-Herald,* October 15, 1921; *Birmingham Post,* December 20, 1921. A clear precedent for this attention to worker needs, and one with which company officials were surely familiar, was the "welfare capitalism" practiced at TCI in Birmingham. See Rikard, "An Experiment in Welfare Capitalism," 91–130.

11. W. R. Lloyd to O. G. Thurlow, September 29, 1921; Dr. S. R. Benedict to W. M. Walmsley, November 25, 1921. These letters and the monthly personnel and casualty reports can be found in the Mitchell Dam File, Alabama Power Company Archives. Interview with L. B. Groover, Jr., February 5, 1995, Montgomery, Alabama. Labor turnover was a big problem in many Alabama industries and especially those in the Birmingham

District. For a discussion of efforts to control this, see Rikard, "George Gordon Crawford," 165–70.

12. *Montgomery Advertiser,* September 18, 1921; surgeon's report for March 1922, Mitchell Dam File, Alabama Power Company Archives; *Powergrams* (March 1922), 17, (September 1921), 4–5.

13. *Clanton Union Banner,* October 6, 1921; *Birmingham Post,* December 20, 1921.

14. *Clanton Union Banner,* September 29, 1921, February 22, 1923; W. R. Lloyd to O. G. Thurlow, September 29, 1921, Mitchell Dam File, Alabama Power Company Archives. The letters between Lawrence F. Gerald and Walter M. Hood (October-November 1921) pertaining to the four men arrested for bootlegging are in the Alabama Power Company legal file in the company archives.

15. *Montgomery Advertiser,* September 18, 1921; L. N. Branch to J. M. Barry, November 3, 1921; Lawrence F. Gerald to Walter M. Hood, October 25, 1921; James M. Barry to Walter M. Hood, November 5, 1921, all in the Legal File, Alabama Power Company Archives.

16. Casualty department report for February 1922, Mitchell Dam File, Alabama Power Company Archives.

17. Casualty department report for September 1921, Mitchell Dam File, Alabama Power Company Archives.

18. *Birmingham News,* September 18, 1921; *Powergrams* (April 1923), 16, 35, (November 1921), 1.

19. *Powergrams* (November 1921), 1–2, 24, (October 1922), 12; *Birmingham News,* September 18, 1921, September 24, 1922.

20. *Powergrams* (September 1921), 4; *Clanton Union Banner,* November 17, 1921.

21. *Powergrams* (November 1921), 1–2, 24, (October 1922), 12.

22. O. G. Thurlow to Maj. J. J. Loving, November 21, 1921, Mitchell Dam Files, Alabama Power Company Archives.

23. *Powergrams* (September 1922), 1, (October 1922), 12–13; Dixie Construction, progress report for March 1922, Mitchell Dam Files, Alabama Power Company Archives.

24. *Powergrams* (September 1922), 1–2, 19, (October 1922), 13, 18–19; *Clanton Union Banner,* December 21, 1922.

25. *Powergrams* (January 1923), 1–2; *Montgomery Advertiser,* June 28, 1921.

26. *Powergrams* (January 1923), 1–2.

27. Ibid.

Chapter Five. Life at Camp Mitchell

1. Interview with Geraldine H. Lawrence, April 26, 1995, Clanton, Alabama; interview with Frank D. Greene, April 26, 1995, Clanton, Alabama.

2. *Lock 12 Development,* 1464–66; *History of Construction of Lock 12,* 35–36.

3. Dr. S. R. Benedict to W. M. Walmsley, November 25, 1921; surgeon's report for January 1922, both in the Mitchell Dam Files, Alabama Power Company Archives. *Powergrams* (May 1922), 1. In addition to being chief surgeon for the Alabama Power Company, Benedict was surgeon for the Seaboard Air Line Railway and local surgeon for the Crane Company and the Casualty Department of the Aetna Life Insurance Company. He also served as attending surgeon to the Hillman and St. Vincents Hospital. Among

his many accomplishments was the development of the "Benedict method of prone pressure resuscitation" for victims of electric shock or drowning, a technique that was soon used throughout the Southeast. See *Powergrams* (September 1922), 3.

4. *Powergrams* (May 1922), 1–2, (November 1923), 14; surgeon's report for January 1922 in the Mitchell Dam Files, Alabama Power Company Archives. See also Rikard, "An Experiment," 131–205.

5. *Powergrams* (September 1921), 4, (March 1922), 14, 18–22, 24, (May 1922), 1.

6. Monthly surgeon's reports are in the Mitchell Dam Files, Alabama Power Company Archives. See report for March 1922.

7. Dr. S. R. Benedict to W. M. Walmsley, November 25, 1921; Dr. S. R. Benedict to W. M. Walmsley, September 22, 1922; surgeon's report for February, April, May 15 to September 30, June, August, September, October 1922, Mitchell Dam Files, Alabama Power Company Archives; *Powergrams* (September 1922), 3, (November 1923), 14–15; *Montgomery Advertiser*, September 18, 1921.

8. These figures were compiled from the surgeon's reports for 1922, Mitchell Dam Files, Alabama Power Company Archives.

9. Casualty department reports for April, June, July 1922; surgeon's report for January, April, May, July, December 1922, all in Mitchell Dam Files, Alabama Power Company Archives. *Clanton Union Banner*, January 26, 1922; *Powergrams* (September 1921), 5, (February 1922), 18.

10. Surgeon's reports for March, April, May, August 1922, in Mitchell Dam Files, Alabama Power Company Archives.

11. W. R. Lloyd to O. G. Thurlow, September 29, 1921, in the casualty department reports, Mitchell Dam Files, Alabama Power Company Archives; *Powergrams* (February 1922), 18.

12. *Powergrams* (February 1922), 18, (March 1922), 6, (September 1922), 5; *Birmingham Age-Herald*, October 15, 1921.

13. *Powergrams* (June 1922), 9, (August 1922), 19, (September 1922), 5, (December 1922), 5; Murray, *Thomas W. Martin*, 64.

14. *Powergrams* (August 1922), 19, (September 1922), 5, (December 1922), 5–6; interview with Geraldine H. Lawrence, April 26, 1995, Clanton, Alabama.

15. *Powergrams* (February 1922), 18, (March 1922), 6, (August 1922), 19, (September 1922), 5; *Clanton Union Banner*, February 2, 1922.

16. *Powergrams* (February 1922), 18; interview with Frank D. Greene, April 26, 1995, Clanton, Alabama; interview with Fred G. Mayfield, April 26, 1995, Clanton, Alabama. Mayfield was hired in part because of his basketball skills.

17. *Chilton County News*, September 21, 1922; *Clanton Union Banner*, March 23, June 29, 1922; interview with John D. Glascock, June 15, 1995, Jemison, Alabama.

18. *Powergrams* (March 1922), 5, (May 1922), 7, (June 1922), 24, (September 1922), 5, (October 1922), 6; *Birmingham Post*, December 20, 1921. See also Robert J. Norrell, *A Promising Field: Engineering at Alabama, 1837–1987* (Tuscaloosa, Ala., 1990), 94–95, 111–12. Visits by engineering students from the University of Alabama and Auburn were regular events, and, according to Norrell (94), Alabama Power "gave preference in hiring to engineering graduates of the university." Indeed, Alabama Power Company was the single largest employer of Alabama engineering graduates for the two decades after World War I.

19. Survey report, January 31, 1922; monthly report for August, 1922, both in the Mitchell Dam Files, Alabama Power Company Archives. *Powergrams* (December 1921), 12, (February 1922), 2, (December 1922), 13, 22.

20. *Powergrams* (January 1922), 18, (June 1922), 102, (February 1923), 12; *Birmingham News*, September 24, 1922.

21. *Powergrams* (February 1923), 12.

22. Ibid., 12–13.

23. *Powergrams* (February 1923), 12, (May 1923), 25, (August 1923), 7; construction report, February and March 1923; surgeon's report, January, February, March 1923, all in Mitchell Dam Files, Alabama Power Company Archives.

24. *Birmingham News*, September 18, 1921; *Powergrams* (June 1922), 17, (August 1922), 5, (April 1923), 38.

25. *Birmingham News*, September 24, 1922; L. C. Sims to Dr. S. R. Benedict, August 8, 1923; Dr. S. R. Benedict to E. A. Yates, September 12, 1923; surgeon's report for July 1923, all in Mitchell Dam Files, Alabama Power Company Archives.

26. *Powergrams* (October 1923), 8–9.

27. *Powergrams* (July 1924), 2, (July 1925), 16.

28. *Powergrams* (April 1923), 1–3; *Montgomery Advertiser*, July 27, 1923. See also J. S. Sutherland to J. M. Barry, December 7, 1923; J. M. Barry to J. U. Benzinger, December 12, 1923; J. U. Benzinger to J. M. Barry, December 17, 1923; and J. M. Barry to J. U. Benzinger, December 22, 1923, all in the Mitchell Dam Files, Alabama Power Company Archives.

Chapter Six. Finally, a Dam at Cherokee Bluffs

1. Interview with John D. Towns, March 9, 1995, Alexander City, Alabama; J. F. Fargason, "History of the Construction of Martin Dam," *The Eclectic Observer*, June 2, 1994; *Tallapoosa News*, August 10, 1923; *Birmingham News*, February 9, 1912; Guesna (Mrs. Wilmot) Moon to Harvey H. Jackson III, February 3, 1995, in possession of the author.

2. *Powergrams* (August 1924), 4–5; *Clanton Union Banner*, July 14, 1921; *Birmingham News*, February 14, 1912.

3. *Powergrams* (August 1924), 4–5; *Clanton Union Banner*, June 2, 1920. The records pertaining to the dam at Cherokee Bluffs are the most complete and best organized of the first four hydroelectric projects. Most of them are contained in bound volumes and are located in the Alabama Power Company Archives. See basin surveys, Martin Dam, Book 4, 2.

4. *Alexander City Outlook*, September 1, 1922.

5. *Birmingham News*, June 8, 1923; *Dadeville Spot Cash*, June 20, 1923; *Alexander City Outlook*, June 13, June 20, 1923; *Tallapoosa News*, June 22, 1923; *Powergrams* (June 1923), 19, (October 1923), 3, (February 1924), 16.

6. *Powergrams* (August 1923), 16; *Dadeville Spot Cash*, June 27, 1923.

7. *Tallapoosa News*, July 6, 1923.

8. *Powergrams* (June 1925), 3.

9. *Eclectic Observer*, June 2, 1994; *Powergrams* (June 1925), 3–4, (July 1925), 4, 9, (November 1925), 25.

10. Martin Dam, Book 4, 38–39.

11. *Alabama Sportsman* (August-September 1926), 2; Frank M. Dunlap to Harvey H. Jackson III, May 20, 1995, in possession of author; *Dadeville Spot Cash*, March 12, 1925; interview with Clyde Steverson, March 20, 1995, Alexander City, Alabama.

12. "Lake Martin," in *Tallapoosa County: A History* (Tallapoosa County Bicentennial Committee, 1976), 151; *Eclectic Observer*, June 2, 1994; interview with John D. Towns, March 9, 1995, Alexander City, Alabama; interview with B. K. McDonald, March 2, 1995, Wetumpka, Alabama.

13. Martin Dam, Book 4, 8; *Powergrams* (August 1923), 7, (September 1923), 15; interview with B. K. McDonald, March 2, 1995, Wetumpka, Alabama; interview with Gordon Gauntt, April 23, 1995, Tallassee, Alabama.

14. According to Emfinger, he was the "first" worker hired at Cherokee Bluffs. Interview with Lloyd Frank Emfinger, April 23, 1995, Tallassee, Alabama; interview with Marguerite "Bill" Henry (Mrs. Thomas) Roach, April 29, 1995, May 23, 1995; interview with B. K. McDonald, March 2, 1995, Wetumpka, Alabama; interview with Ben Hyde, March 10, 1995, Dadeville, Alabama; *Powergrams* (March 1993), 16.

15. Martin Dam, Book 3, 30; interview with Fred G. Mayfield, April 26, 1995, Clanton, Alabama.

16. Martin Dam, Book 3, 31–32; interview with L. B. O'Daniel, March 10, 1995, Tallassee, Alabama.

17. Martin Dam, Book 3, 32.

18. Ibid.

19. Fred G. Mayfield to Harvey H. Jackson III, April 24, 1995, in possession of the author; interview with Fred G. Mayfield, April 26, 1995, Clanton, Alabama; interview with Lloyd Frank Emfinger, April 23, 1995, Tallassee, Alabama; *Powergrams* (July 1994), 19; "Looking Back: Golden Memories of Martin Dam," *Powergrams* (February 1993), 19–21; *Powergrams* (August 1923), 7, (July 1924), 19.

20. Interview with L. B. O'Daniel, March 10, 1995, Tallassee, Alabama; interview with Ben Hyde, March 10, 1995, Dadeville, Alabama; interview with Beula Golden Ingram, March 9, 1995, Alexander City, Alabama.

21. *Powergrams* (July 1923), 14, 19, (June 1925), 4.

22. Martin Dam, Book 3, 45–46, 88.

23. *Powergrams* (August 1923), 17, (September 1923), 14; Martin Dam, Book 3, 31, 49.

24. Martin Dam, Book 3, 46–47.

25. *Powergrams* (August 1923), 17, (September 1923), 14–15; Martin Dam, Book 1, 165–66; Martin Dam, Book 3, 48, 54, 96, 121–23; interview with Dr. J. F. Fargason, April 22, 1995, Tallassee, Alabama.

26. Martin Dam, Book 3, 48–49, 90–91; interview with L. B. O'Daniel, March 10, 1995, Tallassee, Alabama; interview with Beula Golden Ingram, March 9, 1995, Alexander City, Alabama.

27. Martin Dam, Book 3, 50, 91, 96; interview with L. B. O'Daniel, March 10, 1995, Tallassee, Alabama.

28. Martin Dam, Book 3, 46, 51; interview with John D. Towns, March 9, 1995, Alexander City, Alabama; *Eclectic Observer*, June 2, 1995.

29. Martin Dam, Book 3, 49–52; interview with Beula Golden Ingram, March 9, 1995, Alexander City, Alabama.

Chapter Seven. A Dam for Mr. Martin

1. *Alabama Sportsman* (August-September 1926), 2; *Powergrams* (August 1923), 16, (November 1923), 21, (December 1923), 23, (February 1924), 16, (June 1924), 5; Martin Dam, Book 2, 2–3; Martin Dam, Book 3, 72; *Dadeville Spot Cash,* September 12, 1923; interview with L. B. O'Daniel, March 10, 1995, Tallassee, Alabama.

2. Martin Dam, Book 5, 2–4; Frank M. Dunlap to Harvey H. Jackson III, January 31, 1995, in possession of the author; *Alexander City Outlook,* August 8, 1923; *Eclectic Observer,* June 2, 1994; *Powergrams* (June 1924), 5, (July 1925), 5; interview with John D. Towns, March 9, 1995, Alexander City, Alabama; interview with L. B. O'Daniel, March 10, 1995, Tallassee, Alabama.

3. Martin Dam, Book 4, 8, 12; *Powergrams* (July 1923), 9, (August 1923), 17, (October 1923), 3, 6, (June 1924), 5; *Alabama Sportsman* (August-September 1926), 2; interview with B. K. McDonald, March 2, 1995, Wetumpka, Alabama.

4. Martin Dam, Book 5, 3–4; *Powergrams* (June 1924), 5; "Lake Martin," in *Tallapoosa County,* 151; interview with B. K. McDonald, March 2, 1995, Wetumpka, Alabama.

5. *Powergrams* (October 1923), 3, 6, (June 1925), 4, 9; *Eclectic Observer,* June 2, 1994.

6. Martin Dam, Book 4, 39–40; interview with John D. Towns, March 9, 1995, Alexander City, Alabama; *Eclectic Observer,* June 2, 1994.

7. Martin Dam, Book 4, 39–40; *Eclectic Observer,* June 2, 1994; "Lake Martin," in *Tallapoosa County,* 151; Mrs. M. L. Young to Capt. James R. Hall, June 26, 1925, in Martin Dam Files, Alabama Power Company Archives.

8. *Alexander City Outlook,* September 1, 1922, June 20, July 8, 1923; *Montgomery Advertiser,* July 8, 1923; *Birmingham News,* November 8, 1925; *Powergrams* (June 1925), 3.

9. Martin Dam, Book 3, 104–7, 121; *Powergrams* (November 1923), 14–15; "Lake Martin," in *Tallapoosa County,* 151; interview with Beula Golden Ingram, March 9, 1995, Alexander City, Alabama.

10. Martin Dam, Book 3, 105–7; *Powergrams* (November 1923), 14–15, (February 1924), 17, (July 1925), 17.

11. *Powergrams* (February 1924), 17–18; Martin Dam, Book 3, 105; interview with Howard F. Bryan III, February 5, 1995, Montgomery, Alabama; interview with Lloyd Frank Emfinger, April 23, 1995, Tallassee, Alabama; interview with L. B. "Gip" O'Daniel, March 10, 1995, Tallassee, Alabama; interview with B. K. McDonald, March 2, 1995, Wetumpka, Alabama.

12. Martin Dam, Book 3, 47, 50, 107; Martin Dam, Book 4, 29–32; *Powergrams* (November 1923), 18, (July 1925), 17.

13. *Alabama Sportsman* (August-September 1926), 2, 4; *Powergrams* (July 1925), 17, (December 1927), 7; interview with Dr. J. F. Fargason, April 22, 1995, Tallassee, Alabama; interview with John D. Towns, March 9, 1995, Alexander City, Alabama.

14. Interview with John D. Towns, March 9, 1995, Alexander City, Alabama; John D. Towns to Harvey H. Jackson III, May 30, 1995, in possession of the author; *Powergrams* (August 1923), 18, (December 1924), 1, (July 1925), 17, (December 1927), 7–8; Frank M. Dunlap to Harvey H. Jackson III, January 31, 1995, in possession of the author.

15. Interview with Lloyd Frank Emfinger, April 23, 1995, Tallassee, Alabama; *Powergrams* (July 1923), 19, (February 1924), 16, (September 1924), 38.

16. Interview with Dr. J. F. Fargason, April 22, 1995, Tallassee, Alabama.

17. Interview with B. K. McDonald, March 2, 1995, Wetumpka, Alabama; interview with John D. Towns, March 9, 1995, Alexander City, Alabama; interview with L. B. Crouch, Tallassee, Alabama; interview with Gordon Gauntt, April 23, 1995, Tallassee, Alabama.

18. Martin Dam, Book 1, 165–68; Martin Dam, Book 3, 51–53; *Powergrams* (February 1924), 16–17; J. Oran Hardin to Jim Cassidy, February 9, 1995, copy in possession of the author.

19. Martin Dam, Book 1, 174–75.

20. Martin Dam, Book 1, 80, 182; *Powergrams* (February 1924), 17–18; *Alabama Sportsman* (August-September 1926), 2; *Jordan Dam, Construction Report, Narrative,* 126. Typescript volume in Alabama Power Company Archives.

21. Interview with Marguerite "Bill" Henry Roach, April 29, 1995, Grand Bay, Alabama; interview with John D. Towns, March 9, 1995, Alexander City, Alabama; *Powergrams* (August 1923), 18, (December 1927), 8; "Lake Martin," in *Tallapoosa County,* 151.

22. *Powergrams* (August 1924), 4, (November 1925), 1, 25–27; Murray, *Thomas W. Martin,* 74; Jackson, *Rivers of History,* 180; *Dadeville Spot Cash,* October 29, 1925, November 12, 1926.

23. *Powergrams* (July 1926), 2; *Dadeville Spot Cash,* June 24, 1926; *Alexander City Outlook,* June 24, 1926.

24. Martin Dam, Book 3, 30. Stories of Mexican workers were told by many former workers on the Martin Dam project. Interview with Fred G. Mayfield, April 26, 1995, Clanton, Alabama; interview with Clyde Steverson, May 20, 1995, Alexander City, Alabama; interview with Dr. J. F. Fargason, April 22, 1995, Tallassee, Alabama; interview with Ben Hyde, March 10, 1995, Dadeville, Alabama; interview with B. K. McDonald, March 2, 1995, Wetumpka, Alabama. For evidence of Alabama Power's interest in Mexico, see a booklet in the Power Company Archives entitled "Mexican Northern Power Company, Limited," and F. S. Washburn to F. Darlington, June 12, 1995.

25. Martin Dam, Book 4, 51, 67–68; *Powergrams* (October 1926), 28; "Kowaliga Creek Bridge," 1, transcript of article provided by Ben Hyde; *Montgomery Advertiser,* September 19, 1926.

26. Martin Dam, Book 4, 56, 66–67; *Powergrams* (October 1926), 28, (October 1927), 5; *Alabama Sportsman* (August-September 1926), 4; "Kowaliga Creek Bridge," 2; *Birmingham News,* January 3, 1927.

27. *Powergrams* (October 1927), 4–5; Martin Dam, Book 4, 55–56, 67–68; "Kowaliga Creek Bridge," 3–4; interview with Ben Hyde, March 10, 1995, Dadeville, Alabama.

28. Interview with Howard F. Bryan III, February 5, 1995, Montgomery, Alabama.

29. Martin Dam, Book 1, 186; interview with Howard F. Bryan III, February 5, 1995, Montgomery, Alabama.

Chapter Eight. Taming the Devil's Staircase

1. *Jordan Dam, Construction Report,* 4–5; Jackson, *Rivers of History,* 143–45, 149; O. G. Thurlow to Henry F. Carter, February 17, 1915, Lay Dam Files, Alabama Power Company Archives.

2. *Alexander City Outlook,* April 22, 1926; Martin, *Forty Years,* 27.

3. *Powergrams* (August 1925), 9, (October 1927), 1–2; Murray, *Thomas W. Martin,*

78–79. In 1928 *Powergrams* gave Jordan Dam much less coverage than it had the Mitchell and Martin projects. Only September and December issues had major articles.

4. Interview with B. K. McDonald, March 2, 1995, Wetumpka, Alabama. Other interviews, cited below, also mentioned how men came from other sites.

5. Interview with Sarah Cabot Robison (Mrs. Donald) Pierce, March 2, 1995, Montgomery, Alabama; *Powergrams* (August 1925), 9.

6. For examples of the way navigation was worked into many of the articles on the building of Alabama Power Company dams, see *Powergrams* (July 1925), 8, (September 1928), 5. See also Jackson, *Rivers of History*, chapters 11 and 12.

7. *Powergrams* (July 1925), 8, (September 1928), 5. See also Jackson, *Rivers of History*, chapters 12 and 13.

8. Interview with Sarah Cabot Robison Pierce, March 2, 1995, Montgomery, Alabama; interview with Ben Hyde, March 10, 1995, Dadeville, Alabama; *Powergrams* (November 1925), 24.

9. *Jordan Dam, Construction Report*, 142; *Powergrams* (July 1927), 7, (September 1928), 5.

10. Interview with Howard F. Bryan III, February 5, 1995, Montgomery, Alabama.

11. *Jordan Dam, Construction Report*, 33, 82–85; interview with Marguerite "Bill" Henry Roach, April 29, 1995, Grand Bay, Alabama.

12. *Jordan Dam, Construction Report*, 89; interview with L. B. Groover, Jr., February 5, 1995, Montgomery, Alabama.

13. Interview with Howard F. Bryan III, February 5, 1995, Montgomery, Alabama; *Powergrams* (July 1927), 7; *Jordan Dam, Construction Report*, 90.

14. Interview with B. K. McDonald, March 2, 1995, Wetumpka, Alabama; *Powergrams* (March 1993), 16–17.

15. Interview with B. K. McDonald, March 2, 1995, Wetumpka, Alabama.

16. Ibid.; *Jordan Dam, Construction Report*, 84.

17. Interview with Sarah Cabot Robison Pierce, March 2, 1995, Montgomery, Alabama.

18. Interview with B. K. McDonald, March 2, 1995, Wetumpka, Alabama; interview with L. B. Groover, Jr., February 5, 1995, Montgomery, Alabama; interview with Marguerite "Bill" Henry Roach, April 29, 1995, Grand Bay, Alabama. The rumor that the sheriff was involved in bootlegging was confirmed by Harvey H. Jackson, Jr., of Grove Hill, Alabama.

19. Interview with B. K. McDonald, March 2, 1995, Wetumpka, Alabama.

20. *Jordan Dam, Construction Report*, 89–90, 94–95; interview with B. K. McDonald, March 2, 1995, Wetumpka, Alabama.

21. *Powergrams* (December 1927), 6.

22. *Jordan Dam, Construction Report*, 90–91; interview with B. K. McDonald, March 2, 1995, Wetumpka, Alabama.

23. *Jordan Dam, Construction Report*, 143; interview with Marguerite "Bill" Henry Roach, April 29, 1995, Grand Bay, Alabama.

24. Interview with L. B. Groover, Jr., February 5, 1995, Montgomery, Alabama; interview with Howard F. Bryan III, February 5, 1995, Montgomery, Alabama; interview with Marguerite "Bill" Henry Roach, April 29, 1995, Grand Bay, Alabama.

25. Interview with Marguerite "Bill" Henry Roach, April 29, 1995, Grand Bay, Alabama; interview with Howard F. Bryan III, February 5, 1995, Montgomery, Alabama.

26. *Jordan Dam, Construction Report,* 164–68.

27. Interview with B. K. McDonald, March 2, 1995, Wetumpka, Alabama.

28. Material on the dedication can be found in the Jordan Dam Files, Alabama Power Company Archives. See also Murray, *Thomas W. Martin,* 78–80. Interview with B. K. McDonald, March 2, 1995, Wetumpka, Alabama.

29. *Powergrams* (April 1923), 11, (June 1923), 18, (July 1923), 17–18, (October 1923), 15, 17, (April 1927), 32–33; Murray, *Thomas W. Martin,* 80.

30. Murray, *Thomas W. Martin,* 27, 36, 45–47, 67, 81; Martin, *Forty Years,* 27; *Powergrams* (April 1923), 11, (July 1923), 17–18, (October 1923), 12, (March 1924), 17–18, (April 1927); Martin Dam, Book 4, 42–43.

31. Material on the dedication may be found in the Lay Dam Files, Alabama Power Company Archives.

Chapter Nine. After the Dams

1. *Powergrams* (December 1920), 9. One of the most consistent points made by the people interviewed was how good things were in the company villages.

2. *Powergrams* is filled with examples of life in the villages. For an article on Benzinger see August 1920, 19.

3. Interview with Howard F. Bryan III, February 5, 1995, Montgomery, Alabama.

4. Almost every issue of *Powergrams* noted visitors to the various sites. See June 1920, p. 8; December 1920, p. 22; May 1921, p. 9; June 1921, p. 8; October 1921, p. 4.

5. *Powergrams* (April 1927), 5, 24. For ways other companies employed the recreational aspect of welfare capitalism in their operations, see Brandes, *American Welfare Capitalism,* 75–82.

6. *Powergrams* (March 1924), 3, (April 1927), 5, 24.

7. *Powergrams* (September 1923), 8, (March 1924), 1–3, (April 1927), 5, 24.

8. *Powergrams* (June 1923), 6–7, 30, (December 1923) 22, (March 1924), 2, (May 1924), 14.

9. *Powergrams* (August 1922), 5, (March 1924), 1, (June 1924), 16, 32, (July 1925), 31–32.

10. *Powergrams* (April 1923), 28, (June 1923), 7, (August 1923), 3, 14–15, (October 1924), 21. Interview with Geraldine H. Lawrence, April 26, 1995, Clanton, Alabama.

11. *Powergrams* (June 1923), 12, (August 1923), 3, (August 1924), n.p.; *Eclectic Observer,* June 2, 1994; *Alabama Sportsman* (August-September 1926), 3; Martin Dam, Book 3, 50, 122.

12. *Tallapoosa News,* October 21, 1926; Martin Dam, Book 4, 42–44; *Powergrams* (December 1927), 8; *Jordan Dam, Construction Record,* 143. Interview with Dr. J. F. Fargason, April 22, 1995, Tallassee, Alabama. Dr. Fargason was one of the mosquito inspectors at Lake Martin.

13. "Bits of History of Early Electronic Power Plants in Alabama," compiled by Alabama Power for Thomas W. Martin, typescript copy, Alabama Power Company Archives, 1941, preface. Alabama Power Company *Annual Report,* 1928, 3.

222

14. Alabama Power Company *Annual Report*, 1921, n.p., 1926, 4; *Tallapoosa News*, March 26, 1926; *Powergrams* (April 1923), 28, (July 1923), 1–3. Most of the October 1926 issue of *Powergrams* is devoted to rural electrification.

15. Alabama Power Company *Annual Report*, 1923, 9–10, 1924, 21–23, 1926, 7, 1926, 13.

16. Murray, *Thomas W. Martin*, 87–102; interview with B. K. McDonald, March 2, 1995, Wetumpka, Alabama. Most of the other people interviewed observed that those who were able to keep their jobs and live in a company village were much better off than their counterparts.

Bibliographical Essay

Alabama Power Company records pertaining to the early dams are housed in the company's archives in Birmingham, Alabama. They consist of correspondence, financial records, engineering reports and drawings, newspaper clippings, programs from the various dam dedications, and legal documents. Unfortunately, these records are not complete, for not until the Cherokee Bluffs project did the company begin to maintain extensive files on the work being done. Other material, such as the movies made at Mitchell Dam, has been lost but may still be recovered.

State and local newspapers were also extremely helpful. Papers from Birmingham and Montgomery kept the public well informed on the progress being made and on the impact the dams would have on industrial development in Alabama. Local newspapers were, naturally, more parochial in their emphasis, an approach that made their articles essential to understanding the way these projects affected the regions in which they were being built. For a broader picture, see the *Birmingham News,* the *Birmingham Age-Herald,* the *Birmingham Post,* the *Montgomery Advertiser,* and the *Montgomery Journal.* Local accounts were found in the *Clanton Union Banner,* the *Chilton County News,* the *Chronicle* (Rockford, Ala.), the *Alexander City Outlook,* the *Tallapoosa News,* the *Dadeville Spot Cash,* the *Eclectic Observer,* the *Gadsden Evening Journal,* the *Gadsden Times News,* the *Anniston Star,* and the *Wetumpka Herald.*

Important information on Lock 12/Lay Dam came from an unbound typed manuscript entitled *History of Construction of Lock 12: Dam and Foundation: Coosa River,* written by Resident Engineer A. C. Polk and located in the company archives. It is identified as volume 1, but no other volumes have been found. Also useful was E. L. Sayers and A. C. Polk, *The Lock 12 Development of the Alabama Power Company, Coosa River, Alabama,* which was reprinted from the *Transactions* of the American Society of Civil Engineers 77 (1915), 1409–1581. *A Study of Impounded Waters of the Coosa River in Shelby, Chilton, Talladega and Coosa Counties, Alabama,* by U.S. Public Health Service Sanitary Engineer J. A. Le Prince, deals with the health problems that led to the "mosquito suits," as do the legal files in the archives.

Not surprising, the best-kept and -preserved records for Mitchell Dam are the ones dealing with personnel, casualty, and health matters; these are in manuscript in the company archives. Equally important is the appearance of *Powergrams,* which, despite its role in promoting the company, is an accurate and excellent source for the construction of the facility. Because the company was very interested in presenting a "human face" to the public, *Powergrams* contains more social commentary than might be expected in a publication of this sort. The many instances when I used this valuable resource are cited in the footnotes.

Of all the early projects, Martin Dam at Cherokee Bluffs is the best documented. Five bound typescript volumes cover every important aspect of the work, and when supplemented by correspondence and newspaper reports, they give a comprehensive picture of the work done there. In addition *Powergrams*, by publishing what was going on at other sites while Martin Dam was being built, helps to show how the company was developing its entire operation.

Jordan Dam, the last of these early efforts, drew much less attention; thus there are fewer records. One bound, typed volume is kept in the archives, along with a few articles in *Powergrams*. Correspondence pertaining to this dam is also thin, and there are surprisingly few newspaper articles. Apparently the company had learned to avoid controversy, and the only "news" was "old news."

Also helpful in understanding the evolution of the company were its annual reports. The first of these was published for the year 1921, although some of the information contained therein related to earlier years. Taken together, these reports enable one to trace the evolution of the company, especially in its service to its customers. Around 1940, curious about the generation of electricity in the state prior to 1912, Thomas Martin commissioned a study to reveal what existed when Lay Dam came on line. The result can be found in a typescript volume entitled "Bits of History of Early Electronic Power Plants in Alabama," compiled by Alabama Power for Thomas W. Martin, 1941, which is located in the company archives.

Although most of the material for this study came from the Alabama Power Company Archives, a number of other books and articles proved valuable. For a general account of river development, see Harvey H. Jackson III, *Rivers of History: Life on the Coosa, Tallapoosa, Cahaba, and Alabama* (Tuscaloosa, Ala., 1995). Specific information on the early days of Alabama Power can be found in Thomas W. Martin, *Forty Years of Alabama Power Company* (New York, 1952); Thomas W. Martin, *The Story of Electricity in Alabama since the Turn of the Century* (Birmingham, 1952); and William M. Murray, Jr., *Thomas W. Martin: A Biography* (Birmingham, 1978). Though all present an uncritical look at their subjects, they are still valuable. See also John R. Hornaday, *Soldiers of Progress and Industry* (New York, 1930); R. A. Mitchell, *S. Z. Mitchell and the Electrical Industry* (New York, 1960); and Hughes Reynolds, *The Coosa River Valley from De Soto to Hydroelectric Power* (Cynthiana, Ky., 1944).

Valuable in understanding the context in which Alabama Power operated and the possible source for many of its ideas on how labor camps should be organized was Marlene Hunt Rikard, "An Experiment in Welfare Capitalism: The Health Care Services of the Tennessee Coal, Iron and Railroad Company" (Ph.D. diss., University of Alabama, 1983); and Marlene Hunt Rikard, "George

Gordon Crawford: Man of the New South," *Alabama Review* 31 (July 1978): 163–81. Also helpful were James Seay Brown, Jr., *Up Before Daylight: Life Histories from the Alabama Writers' Project, 1938–1939* (Tuscaloosa, Ala., 1982); Robert J. Norrell, *A Promising Field: Engineering at Alabama, 1837–1987* (Tuscaloosa, Ala., 1990); Henry M. McKiven, Jr., *Iron & Steel: Class, Race, and Community in Birmingham, Alabama, 1875–1920* (Chapel Hill, N.C., 1995); W. David Lewis, *Sloss Furnaces and the Rise of the Birmingham District: An Industrial Epic* (Tuscaloosa, Ala., 1994), and William Warren Rogers, et al., *Alabama: The History of a Deep South State* (Tuscaloosa, Ala., 1994). For insight into "welfare capitalism" and its role in industrial development early in the century, see Stuart D. Brandes, *American Welfare Capitalism, 1880–1940* (Chicago, 1976); and Gerald Zahavi, *Workers, Managers, and Welfare Capitalism: The Shoeworkers and Tanners of Endicott Johnson, 1890–1950* (Urbana, 1988).

Apart from the many articles in *Powergrams,* the following were also helpful: J. F. Fargason, "History of the Construction of Martin Dam," *The Eclectic Observer,* June 2, 1994; "Souvenir Issue of Lake Martin, the Largest Artificial Lake in the United States, and Martin Dam, Which Impounds This Great Body of Water Created by the Alabama Power Company," *Alabama Sportsman* 2, 9 (August-September 1929); and "Lake Martin," in *Tallapoosa County: A History* (Tallapoosa County Bicentennial Committee, 1976).

Interviews with individuals who could recall the dams being built and the impact these projects had on people in the area were essential to this study, and those involved are acknowledged elsewhere. The tapes of these interviews and transcripts of the conversations, along with correspondence pertaining to the research, have been deposited in the archives of the Alabama Power Company. Copies of this material have also been put in the Oral History Collection at Jacksonville State University, Jacksonville, Alabama.

Index

About the Author

Harvey H. Jackson III is Head of the History Department at Jacksonville State University, Alabama. He has written many works, including *Rivers of History: Life on the Coosa, Tallapoosa, Cahaba, and Alabama* (1995). He received his master's degree from The University of Alabama and his doctorate from the University of Georgia.